INTELLIGENT AUTOMATIC GENERATION CONTROL

INTELLIGENT AUTOMATIC GENERATION CONTROL

HASSAN BEVRANI
University of Kurdistan
Kumamoto University

TAKASHI HIYAMA
Kumamoto University

CRC Press
Taylor & Francis Group
Boca Raton London New York

CRC Press is an imprint of the
Taylor & Francis Group, an **informa** business

Cover photos by Robert Kalinowski.

CRC Press
Taylor & Francis Group
6000 Broken Sound Parkway NW, Suite 300
Boca Raton, FL 33487-2742

First issued in paperback 2017

© 2011 by Taylor and Francis Group, LLC
CRC Press is an imprint of Taylor & Francis Group, an Informa business

No claim to original U.S. Government works

ISBN 13: 978-1-138-07623-5 (pbk)
ISBN 13: 978-1-4398-4953-8 (hbk)

**Visit the Taylor & Francis Web site at
http://www.taylorandfrancis.com**

**and the CRC Press Web site at
http://www.crcpress.com**

To Sabah, Bina, and Zana

and

To Junko, Satoko, Masaki, Atsushi, and Fuyuko

Contents

Preface

Automatic generation control (AGC) is one of the important control problems in interconnected power system design and operation, and is becoming more significant today due to the increasing size, changing structure, emerging renewable energy sources and new uncertainties, environmental constraints, and complexity of power systems. Automatic generation control markets require increased intelligence and flexibility to ensure that they are capable of maintaining a generation-load balance, following serious disturbances.

The AGC systems of tomorrow, which should handle complex, multiobjective regulation optimization problems characterized by a high degree of diversification in policies, control strategies, and wide distribution in demand and supply sources, surely must be intelligent. The core of such intelligent systems should be based on flexible intelligent algorithms, advanced information technology, and fast communication devices. The intelligent automatic generation control interacting with other ancillary services and energy markets will be able to contribute to upcoming challenges of future power systems control and operation. This issue will be performed by intelligent meters and data analyzers using advanced computational methods and hardware technologies in both load and generation sides.

Intelligent automatic generation control provides a thorough understanding of the fundamentals of power system AGC, and addresses several new schemes using intelligent control methodologies for simultaneous minimization of system frequency deviation and tie-line power changes to match total generation and load demand, which is required for successful operation of interconnected power systems. The physical and engineering aspects have been fully considered, and most proposed control strategies are examined by real-time simulations.

The present book could be useful for engineers and operators in power system planning and operation, as well as academic researchers and university students in electrical engineering. This book is organized into twelve chapters.

Chapter 1 provides a review on intelligent power system operation and control, and is mainly focused on the application examples of intelligent technologies in Japanese power system utilities. The chapter presents the state of the art of intelligent techniques in Japanese utilities based on the investigation by the Subcommittee of the Intelligent Systems Implementations in Power Systems of Japan. The current situation of intelligent methods application in Japanese power systems in general is described. Artificial intelligent applications in power system planning and control/restoration are addressed, and next steps and future implementations are explained.

Chapter 2 presents the fundamentals of AGC, providing structure, definitions, and basic concepts. The AGC mechanism in an interconnected power system, and the major functions, constraints, and characteristics are described. The role of AGC systems in connection with the power system monitoring/control master stations, and remote site control centers to manage the electric energy, is emphasized. Power system operations and frequency control in different ranges of frequency deviation are briefly explained. A frequency response model is described, and its usefulness for the sake of AGC dynamic analysis and simulation is examined.

Chapter 3 emphasizes the application of intelligent techniques on the AGC synthesis and addresses the basic control configurations with recent achievements. New challenges and key issues concerning system restructuring and integration of distributed generators and renewable energy sources (RESs) are also discussed. The applications of fuzzy logic, neural networks, genetic algorithms, multiagent systems, combined intelligent techniques, and evolutionary optimization approaches on the AGC synthesis problem are briefly reviewed. An introduction to AGC design in deregulated environments is given, and AGC analysis and synthesis in the presence of RESs and microgrids, including literature review, present worldwide status, impacts, and technical challenges, are presented. Finally, a discussion on the future works and research needs is given.

Chapter 4 reviews the main structures, configurations, and characteristics of AGC systems in a deregulated environment and addresses the control area concept in restructured power systems. Modern AGC structures and topologies are described, and a brief description on AGC markets is given. Concepts such as AGC market and market operator, and the need for intelligent AGC markets in the future are also explained. The chapter emphasizes that the new challenges will require some adaptations of the current AGC strategies to satisfy the general needs of different market organizations and the specific characteristics of each power system. The existing market-based AGC configurations are discussed, and an updated frequency response model for decentralized AGC markets is introduced.

Chapter 5 describes a methodology for AGC design using neural networks in a restructured power system. Design strategy includes enough flexibility to set a desired level of performance. The proposed control method is applied to single- and three-control area examples under a bilateral AGC scheme. It is recognized that the learning of both connection weights and neuron function parameters increases the power of learning algorithms, keeping high capability in the training process. It is shown that the flexible neural-network-based supplementary frequency controllers give better area control error minimization and a proper convergence to the desired trajectory than do the traditional neural networks.

Chapter 6 covers the AGC system and related issues concerning the integration of new RESs in the power systems. The impact of power fluctuation produced by variable renewable sources (such as wind and solar units) on the system frequency performance is presented. An updated power system frequency response model for AGC analysis considering RESs and associated issues is introduced. Some nonlinear time-domain simulations on standard power system examples are presented to show that the simulated results agree with those predicted analytically. Emergency frequency control concerning the RESs is discussed. Finally, the need for revising frequency performance standards, further research, and new AGC perspectives is emphasized.

Chapter 7 addresses the application of multiagent systems in AGC design for multiarea power systems. General frameworks for agent-based control systems based upon the foundations of agent theory are discussed. A new multiagent AGC scheme has been introduced. The capability of reinforcement learning in the proposed AGC strategy is examined, and the application of genetic algorithms (GAs) to determine actions and states during the learning process is discussed. The possibility for building more agents, such as estimator agents to cope with real-world AGC systems, is explained. Finally, the proposed methodology is examined on some power system examples. The application results show that the proposed multiagent control schemes provide a desirable performance, even in comparison to recently developed robust control design.

Chapter 8 proposes a Bayesian-network-based multiagent AGC framework, including two agents in each control area for estimating the amount of power imbalance and providing an appropriate control action signal according to load disturbances and tie-line power changes. The Bayesian network's construction, concepts, and parameter learning are explained. Some nonlinear simulations on a standard test system concerning the integration of wind power units, and also a real-time laboratory experience, are performed. The results show the proposed AGC scheme guarantees optimal performance for a wide range of operating conditions.

Chapter 9 gives an overview on fuzzy-logic-based AGC systems with different configurations. Two fuzzy-logic-based AGC design methodologies based on polar information and particle swarm optimization are presented for the frequency and tie-line power regulation in multiarea power systems. By using the proposed polar-information-based fuzzy logic AGC scheme, the megawatt hour (MWh) constraint is satisfied to avoid the MWh contract violation. The particle-swarm-optimization-based fuzzy logic AGC design is used for frequency and tie-line power regulation in the presence of wind turbines. The efficiency of the proposed control schemes is demonstrated through nonlinear simulations.

Chapter 10 presents a coordinated frequency regulation for the small-sized, high-power energy capacitor system and the conventional AGC participating units to improve the frequency regulation performance. To prevent unnecessary excessive control action, two types of restrictions are proposed for the upper and lower limits of the control signal, as well as for the area control error. By the proposed coordination, the frequency regulation performance is highly improved.

Chapter 11 starts by introducing GAs and their applications in control systems. Then, several methodologies are presented for a GA-based AGC design problem: optimal tuning of conventional AGC systems, AGC formulation through a multiobjective GA optimization problem, GA-based AGC synthesis to track the well-known standard robust performance indices, and using GA to improve the learning performance in the AGC systems. The proposed design methodologies are illustrated by suitable examples. In most cases, the results are compared with recently developed robust control designs.

Chapter 12 presents an intelligent multiagent-based AGC scheme for a power system with dispersed power sources such as photovoltaic, wind generation, diesel generation, and energy capacitor system. In the proposed AGC scheme, the energy capacitor system provides the main function of AGC, while the available diesel units provide a supplementary function of the AGC system. A coordination system between the energy capacitor system and the diesel units is proposed. The developed multiagent system consists of three types of agents: monitoring agents for the distribution of required information through a secure computer network; control agents for the charging/discharging operation on the energy storage device, as well as control of regulation power produced by diesel units; and finally, a supervisor agent for the coordination purpose. Experimental studies in a power system laboratory are performed to demonstrate the efficiency of the proposed AGC scheme.

Hassan Bevrani
University of Kurdistan
Kumamoto University

Takashi Hiyama
Kumamoto University

Acknowledgments

Most of the contributions, outcomes, and insight presented in this book were achieved through long-term teaching and research conducted by the authors and their research groups on intelligent control and power system automatic generation control over the years. It is pleasure to acknowledge the support and awards the authors received from various sources, including Kumamoto University (Japan), University of Kurdistan (Iran), Frontier Technology for Electrical Energy (Japan), Kyushu Electric Power Company (Japan), and West Regional Electric Company (Iran).

The authors thank their postgraduate students F. Daneshfar, P. R. Daneshmand, A. G. Tikdari, H. Golpira, Y. Yoshimuta, G. Okabe, and H. Esaki, and their office secretary Y. Uemura for their active role and continuous support. The authors appreciate the assistance provided by Professor Hussein Beurani from University of Tabriz. Finally, the authors offer their deepest personal gratitude to their families for their patience during preparation of this book.

1

Intelligent Power System Operation and Control: Japan Case Study

In the last twenty years, intelligent systems applications have received increasing attention in various areas of power systems, such as operation, planning, control, and management. Numerous research papers indicate the applicability of intelligent systems to power systems. While many of these systems are still under investigation, there already exist a number of practical implementations of intelligent systems in many countries across the world, such as in Japanese electric utilities.[1] In conventional schemes, power system operation, planning, control, and management are based on human experience and mathematical models to find solutions; however, power systems have many uncertainties in practice. Namely, those mathematical models provide only for specific situations of the power systems under respective assumptions. With these assumptions, the solutions of power systems problems are not trivial. Therefore, there exist some limitations for the mathematical-model-based schemes. In order to overcome these limitations, applications of intelligent technologies such as knowledge-based expert systems, fuzzy systems, artificial neural networks, genetic algorithms, Tabu search, and other intelligent technologies have been investigated in a wide area of power system problems to provide a reliable and high-quality power supply at minimum cost. In addition, recent research works indicate that more emphasis has been put on the combined usage of intelligent technologies for further improvement of the operation, control, and management of power systems.

Several surveys on worldwide application of intelligent methodologies on power systems have been recently published.[2,3] Considering the experience backgrounds of the authors, the present chapter is focused on the application examples of intelligent technologies in Japanese power system utilities. This chapter is organized as follows: The current situation of intelligent methods application in the Japanese power system in general is described in Section 1.1. Artificial intelligent applications in power system planning and control/restoration are addressed in Sections 1.2 and 1.3, respectively. Next steps and future implementation are explained in Section 1.4, and finally, the present chapter is summarized in Section 1.5.

1.1 Application of Intelligent Methods to Power Systems

Intelligent systems are currently utilized in Japanese utilities as well as other developed countries. Table 1.1 shows the application areas of intelligent technologies in Japanese power system utilities. Many applications have been proposed in literature in those areas to demonstrate the advantages of intelligent systems over conventional systems. A certain number of actual implementations of intelligent systems is already working toward better and more reliable solutions for problems in power systems. Table 1.2 shows the

TABLE 1.1

Areas of Intelligent Systems Applications

Area of Application	Number of Utilities
Planning of system expansion	1
Load forecasting	4
Reconfiguration of power systems	3
Unit commitment	2
Fault detection and diagnosis	3
Stabilization control and economic load dispatch (ELD)	3
Restoration after fault and simulators for training	4

TABLE 1.2

Applied Intelligent Techniques

Area of Application	Intelligent Technique
Planning of system expansion	Expert system Genetic algorithm Tabu search
Load forecasting	Neural network Fuzzy inference
Reconfiguration of power systems	Expert system Genetic algorithm Tabu search
Unit commitment	Tabu search Fault detection
Diagnosis	Expert systems Neural network
Stabilization control	Fuzzy logic control
Restoration after fault and simulators for training	Database system Expert system Fuzzy inference Genetic algorithm Tabu search

TABLE 1.3

Objectives of Intelligent Systems Applications

Planning
Expression of uncertainties
Achievement of flexible and robust planning
Multiobjective coordination

Operation
Expression of uncertainties
Expression of experience-based rules
Expression of probability
Reduction of computation time

Load Forecasting
Improvement of accuracy

Power System Control
Improvement of control performance and robustness
Expression of nonlinearities
Implementation of experts' experience
Multiobjective coordination

intelligent technologies actually implemented in Japanese utilities for the area of applications shown in Table 1.1.

Many applications of other intelligent technologies, such as multiagent systems, simulated annealing, and data mining, have been proposed in literature in those areas of applications. However, up to now, those technologies have not been implemented in actual power systems. The purpose of the intelligent systems applications is classified into several categories, as shown in Table 1.3.

1.2 Application to Power System Planning

1.2.1 Expansion Planning of Distribution Systems

The implemented intelligent systems for expansion planning of distribution systems have been mostly utilized to achieve the following objectives, shown in Table 1.4. The systems are connected to the distribution automation system through the local area computer networks, and the decisions are made by the intelligent systems for reconfiguration of distribution networks and the removal of unnecessary equipment after getting required data from the specified distribution automation system.

After implementation of an intelligent system, the investment for system expansion is reduced because of the efficient utilization of already existing devices and equipment. Detailed and fast evaluations are available for future

TABLE 1.4

Functions of Expansion Planning System and Utilized Intelligent Technologies

Function	Intelligent Technique
Optimization of voltage and current levels	Multiagent Tabu search (Required time: 5 minutes)
Minimization of distribution loss	Multiagent Tabu search (Required time: 3 minutes)
Optimal location of devices and equipment	Genetic algorithm (Required time: 3 to 30 minutes)
Usability evaluation for associated devices and equipment	(Required time: 10 seconds to 2 hours)
Simulation of restoration	Multiagent Tabu search (Required Time: 30 minutes)
Optimization of distribution system through integration of above functions	(Required time: 1 hour)

expansions following the increase the power demand. Loss minimization of the distribution systems has been achieved as the result of the averaging of transformer usages.

1.2.2 Load Forecasting

The intelligent load forecasting system, actually implemented in one of the Japanese utilities, is composed of several neural networks for considering the changes of load configurations, depending on the seasons, such as spring, spring to summer, summer, summer to autumn, autumn, and winter. For the training of each neural network, the records during five years have been utilized, including the maximum power demand, hourly power demand, and weather data. The forecasting procedure of the total power demand is divided into two steps as follows:

1. Maximum demand is predicted by the system one day ahead based on the weather forecast, including the highest temperature and lowest temperature. The maximum power demands one day before and one week before are also utilized to estimate the maximum power demand of one day ahead.
2. Based on the maximum power demand in step 1, the hourly power demand is predicted.

By the system, weekly load forecasting is also available. The total power demand is also predicted by the neural networks from the individual load forecasting: light loads such as residential loads and heavy loads such as manufacturing companies and commercial areas. The power loss is also estimated by the neural network. In total, thirty-six neural networks are

TABLE 1.5

Estimation Error for Daily Power Demand

Weak Days								
Data	**Error**	**April**	**May**	**June**	**July**	**Aug**	**Sept**	**Yearly**
Weather forecast	Averaged error	1.43	1.12	1.45	2.57	1.99	1.98	1.74
Past records	Averaged error	0.95	0.85	0.81	1.55	1.49	1.36	1.15
All Days								
Data	**Error**	**April**	**May**	**June**	**July**	**Aug**	**Sept**	**Yearly**
Weather forecast	Averaged error	1.59	1.54	1.71	2.70	2.44	2.20	2.02
Past records	Averaged error	1.08	1.30	1.00	1.80	2.26	1.60	1.48

separately utilized for the detailed load forecasting for the light loads and heavy loads, for different times of 11 a.m., 2 p.m., and 7 p.m., and the seasons. The above procedure can be summarized in the following two steps:

1. Power demand is predicted separately for the light loads and heavy loads at 11 a.m., 2 p.m., and 7 p.m. to have the total power demand of one day ahead.
2. Based on the estimated power demands in step 1, the hourly power demand is predicted by another neural network.

Here, it must be noted that correction of the forecasting by human operators is also available on the implemented load forecasting system.

Table 1.5 shows the accuracy of the load forecasting achieved by the implemented intelligent system based on the neural networks. As shown in the table, the precision of the estimation of power demand is relatively high.

1.2.3 Unit Commitment

For the unit commitment of a group of hydro generators along a river in Japan, an intelligent system has been implemented. The utilized technologies include a rule-based system and intelligent searching scheme. The optimal unit commitment for the grouped hydro generators can be determined based on the water level at each dam, the estimated water flow rate, and so on. All the hydro units should be operated especially during the period when the high fuel cost is expected for the thermal generators. The intelligent system has the functions shown in Table 1.6.

Following the scheduling determined by the intelligent system, economic operation of the hydro units has been achieved. It has been shown through the actual operation of the implemented intelligent system that the scheduling given by it is proper, acceptable, and economical. Further improvement of the system performance and additional rules for operational constraints will be implemented on the current intelligent system.

TABLE 1.6

Types of Scheduling of Hybrid Unit

Operation scheduling one day ahead for each hydro unit
Modification of scheduling following actual water levels
Weekly operation scheduling
Investigations and evaluations for simulation
Investigation of operation planning
Evaluation of actual operation
Training
Scheduling considering maintenance
Data maintenance for operation system for hydro units

1.2.4 Maintenance Scheduling

For the optimization of the maintenance scheduling with 4,500 cases yearly
and 600 cases monthly at maximum, the implementation of intelligent technologies such as Tabu search, mathematical planning, neural networks,
simulated annealing, and others were investigated. Tabu search was finally
selected for this purpose. After implementation of the maintenance scheduling system, a certain amount of cost reduction becomes possible and the
required time for the maintenance scheduling is shortened.

1.3 Application to Power System Control and Restoration

1.3.1 Fault Diagnosis

An intelligent system for fault diagnosis has also been operated to support
the human operators in Japan. The system is composed of two parts: the first
part utilizes the rule-based expert system for the classification of the fault
types, and the second part utilizes several neural networks associated with
the fault types to give their probabilities. The faults are classified into six
types, as shown in Table 1.7.

The accuracy of the identification of fault types is shown in Table 1.8. The
accuracy is relatively low for the faults, including sleet jump, galloping, and
grounding through the construction machines because of the shortage of the
faults of those types.

1.3.2 Restoration

An expert-system-based restoration system was installed more than fifteen
years ago in Japan, for the automation of the 110 and 220 kV systems. This
was one of the earliest implementations of intelligent technologies. Plenty of
simulations were performed to acquire knowledge for the expert systems.

TABLE 1.7

Types of Faults

Lightning
Grounding through construction machine
Grounding through tree branches
Shorting through small animals
Galloping
Sleet jump

TABLE 1.8

Accuracy of Fault Identification

Type of Fault	Accurate Identification	Inaccurate Identification	Accuracy (%)
Lightning	65	0	100
Construction machine	3	2	60
Tree branch	15	0	100
Small animal	10	3	77
Galloping	2	3	50
Sleet jump	6	4	60
Total	101	11	90

For the renewal of the system, modifications of programs and the knowledge base, and also the renewal of the computer system, are inevitable. Currently, the system is not operated as an actual operation. However, the system is utilized for the training of operators.

1.3.3 Stabilization Control

Over the years, many optimization methodologies, robust techniques, and expert and intelligent systems have been used to stabilize power systems, and to improve the control performance and operational functions of power utilities during normal and abnormal conditions.[4] Among these methodologies, fuzzy logic systems have practically attracted more attention. One of the specific features of the fuzzy logic power system stabilizers is their robustness as they provide a wider stable region even for the fixed fuzzy control parameters. Application of fuzzy logic controllers has been proposed mainly for power system stabilizers, and a prototype of a personal-computer-based fuzzy logic power system stabilizer (PSS) was placed in service on a hydro unit in June 1997. The prototype was replaced by a fuzzy logic PSS made by a manufacturer in May 1999.[5] The PSS has been working quite well for nearly ten years, including the PC-based prototype stage. Many other applications have also been proposed in the literature, as shown in Table 1.9, but only a few cases have been implemented.

TABLE 1.9

Application of Fuzzy Logic Controllers

Damping control of oscillations
Control of generators including PSS
Control of FACTS devices
Voltage and reactive power control
Automatic generation control
Others

TABLE 1.10

Problems for Future Extension of Intelligent
Systems in Real Power Systems

Amount of additional investment
Cost of maintenance
Unsatisfactory performance
Required processing speed
Shortage of actual operation
Black-box-based operation
Accuracy of solutions
Acceptability of solutions by human operator

1.4 Future Implementations

As shown in the former sections, there already exist a number of implementations of intelligent systems in Japanese utilities. Some of them are now at their renewal stages. However, because of the reasons listed in Table 1.10, the renewals of some of the intelligent systems will be postponed. Most of these obstacles will be solved by further developments of software/hardware technologies.

Currently, the power system operation and control in all aspects, including automatic generation control (AGC), which is the subject of this book, are undergoing fundamental changes due to restructuring, expanding of functionality, the rapidly increasing amount of renewable energy sources, and the emerging of new types of power generation and consumption technologies. This issue opens the way to realize new/powerful intelligent techniques. The infrastructure of the future intelligent power system should effectively support the provision of ancillary services such as an AGC system from various sources through intelligent schemes.

1.5 Summary

This chapter presents the state of the art of intelligent techniques in Japanese utilities based on the investigation of the Subcommittee of the Intelligent Systems Implementations in Power Systems of Japan. The investigation has been completed, and an investigation of some areas, including the area of automatic generation control, where the actual implementation of intelligent techniques will be expected in the future, has been started.

The rest of this book is focused on the intelligent automatic generation control issue, and several intelligent control strategies are developed for simultaneous minimization of system frequency deviation and tie-line power changes to match total generation and load demand, which is required for the successful operation of interconnected power systems.

References

1. T. Hiyama. 2005. State-of-the art intelligent techniques in Japanese utilities. In *Proceedings of International Conference on Intelligent Systems Application to Power Systems—ISAP*, Arlington, VA, 425–28.
2. Z. Vale, G. K. Venayagamoorthy, J. Ferreira, H. Morais. 2010. Computational intelligence applications for future power systems. In *Computational Intelligence for Engineering Systems*, 180–97. New York: Springer.
3. M. Fozdar, C. M. Arora, V. R. Gottipati. 2007. Recent trends in intelligent techniques to power systems. In *Proceedings of 42nd International Universities Power Engineering Conference (UPEC)*, Brighton, UK, 580–91.
4. H. Bevrani. 2009. Power system control: An overview. In *Robust power system frequency control*, 1–13. New York: Springer.
5. T. Hiyama, S. Oniki, H. Nagashima. 1996. Evaluation of advanced fuzzy logic PSS on analog network simulator and actual installation on hydro generators. *IEEE Trans. Energy Conversion* 11(1):125–31.

2

Automatic Generation Control (AGC): Fundamentals and Concepts

Automatic generation control (AGC) is a significant control process that operates constantly to balance the generation and load in power systems at a minimum cost. The AGC system is responsible for frequency control and power interchange, as well as economic dispatch.

This chapter presents the fundamentals of AGC, providing structure, definitions, and basic concepts. The AGC mechanism in an interconnected power system, the major functions, and characteristics are described. The role of the AGC system in connection with the power system monitoring/control master stations, and remote site control centers to manage the electric energy, is emphasized. Power system operations and frequency control in different ranges of frequency deviation are briefly explained. A frequency response model is described, and its usefulness for the purpose of simulation and AGC dynamic analysis is examined.

2.1 AGC in a Modern Power System

AGC provides an effective mechanism for adjusting the generation to minimize frequency deviation and regulate tie-line power flows. The AGC system realizes generation changes by sending signals to the under-control generating units. The AGC performance is highly dependent on how those generating units respond to the commands.[1] The generating unit response characteristics are dependent on many factors, such as type of unit, fuel, control strategy, and operating point.

The AGC, security control, supervisory control and data acquisition (SCADA), and load management are the major units in the application layer of a modern energy management system (EMS).[2] The AGC process is performed in a control center remote from generating plants, while the power production is controlled by turbine-governors at the generation site. The AGC communicates with SCADA, the load management unit, and the security control center in the EMS, as shown in Figure 2.1.

The SCADA system consists of a master station to communicate with remote terminal units (RTUs) and intelligent electronic devices (IEDs) for a

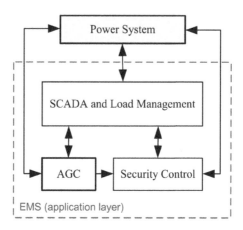

FIGURE 2.1
Application layer of a modern EMS.

wide range of monitoring and control processes. In a modern SCADA system, the monitoring, processing, and control functions are distributed among various servers and computers that communicate in the control center using a real-time local area network (LAN). A simplified SCADA center is shown in Figure 2.2. Although nowadays many data processing and control functions are moved to the IEDs, the power systems still need a master station or control center to organize and coordinate various applications.

As shown in Figure 2.2, the human machine interface (HMI), application servers, and communication servers are the major elements of the SCADA system. The HMI consists of a multi-video-display (multi-VD) interface and a large display or map board/mimic board to display an overview of the power system. The application servers are used for a general database, a historical database, data processing, real-time control functions, EMS configuration, and system maintenance. The communication servers are used for data acquisition from RTUs/IEDs, and data exchange with other control centers.

The data communication, system monitoring, alarms detection, and control commands transmission are the common actions in a SCADA center. Moreover, the SCADA system performs load shedding and special control schemes in cooperation with the AGC system and security control unit. Various security methods and physical options can be applied to protect SCADA systems. To improve the operation security, usually a dual configuration for the operating computers/devices and networks in the form of primary and standby is used.

In a modern SCADA station, the performed control and monitoring processes are highly distributed among several servers, monitors, and communication devices. Using a distributed structure has many advantages, such as easy upgrading of hardware/software parts, reducing costs, and limiting the failures effect. The SCADA system uses open architecture for

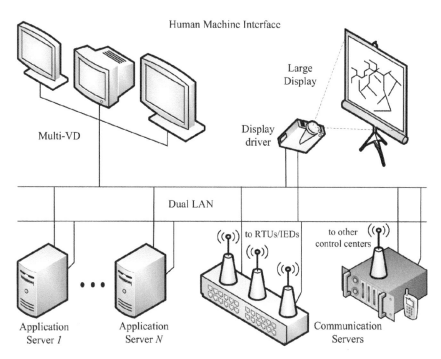

FIGURE 2.2
A simplified structure for a typical SCADA center.

communication with other systems, and to support interfaces with various vendors' products.[3] A mix of communication technologies, such as wireless, fiber optics, and power line communications, could be a viable solution in a SCADA system.

In many power systems, modern communications are already being installed. Substations at both transmission and distribution levels are being equipped with advanced measurement and protection devices as well as new SCADA systems for supervision and control. Communication between control units is also being modernized, as is the communication between several subsystems of the high-level control at large power producers at the EMS level. These are often based on open protocols, notably the IEC61850 family for SCADA-level communication with substations and distributed generating units, and the IEC61968/61970 CIM family for EMS-level communication between control centers.[4]

In some cases, the role of the SCADA system is distributed between several area operating centers; usually one of them is the coordinator and works as the master SCADA center. A real view of an area operating center is shown in Figure 2.3.

In real-power system structures, the AGC centers closely work with the SCADA systems. In this case, a unique SCADA/AGC station effectively uses

FIGURE 2.3
West Area Operating Center at West Regional Electric Co., Kermanshah.

IEDs for doing remote monitoring and control actions. The IEDs as a monitoring and control interface to the power system equipment can be installed in remote (site/substation) control centers and integrated using suitable communication networks. This accomplishes a remote site control system similar to the major station in the SCADA/AGC center. A simplified architecture is presented in Figure 2.4.

A remote site control center may consist of RTU, IEDs, an HMI database server, and a synchronizing time generator. The RTU and IEDs are for communication with the SCADA station, remote access control functions, data measurement/concentration, and status monitoring. The synchronizing time generator is typically a GPS satellite clock that distributes a time signal to the IEDs.

The local access to the IEDs and the local communication can be accomplished over a LAN. Whereas the remote site control center is connected to the SCADA/AGC center, EMS and other engineering systems are through the

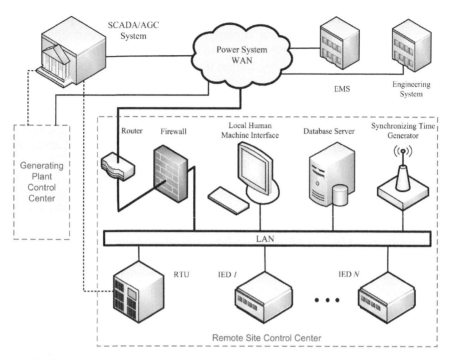

FIGURE 2.4
A simplified architecture including remote site/generating plant controls and SCADA/AGC system.

power system wide area network (WAN). Figure 2.4 shows that the SCADA/AGC center, in addition to the use of WAN (in cooperation with the EMS), can be directly connected to the remote site and generating plant control centers.

Interested readers can find appropriate standards for SCADA systems, substation automation, remote site controls with detailed architectures, and functions of various servers, networks, and communication devices in IEEE PES.[3] The AGC performs a continuous real-time operation to adjust the power system generation to track the load changes economically. Frequency control, economic dispatch, interchange transaction scheduling, reserve monitoring, and related data recording are the main functions of an AGC system, of which frequency control is the most important issue.

2.2 Power System Frequency Control

Frequency deviation is a direct result of the imbalance between the electrical load and the power supplied by the connected generators, so it provides a useful index to indicate the generation and load imbalance. A permanent off-normal frequency deviation directly affects power system operation, security, reliability, and efficiency by damaging equipment, degrading load performance, overloading transmission lines, and triggering the protection devices.

Since the frequency generated in the electric network is proportional to the rotation speed of the generator, the problem of frequency control may be directly translated into a speed control problem of the turbine generator unit. This is initially overcome by adding a governing mechanism that senses the machine speed, and adjusts the input valve to change the mechanical power output to track the load change and to restore frequency to a nominal value.

Depending on the frequency deviation range, as shown in Figure 2.5, in addition to the natural governor response known as the *primary control*, the *supplementary control* (AGC), or secondary control, and *emergency control* may all be required to maintain power system frequency.[1] In Figure 2.5, the f_0 is nominal frequency, and Δf_1, Δf_2, and Δf_3 show frequency variation range corresponding to the different operating conditions based on the accepted frequency operating standards.

Under normal operation, the small frequency deviations can be attenuated by the primary control. For larger frequency deviation (off-normal operation), according to the available amount of power reserve, the AGC is responsible for restoring system frequency. However, for a serious load-generation imbalance associated with rapid frequency changes following a significant fault, the AGC system may be unable to restore frequency via the supplementary frequency control loop. In this situation, the emergency control and protection schemes, such as under-frequency load shedding (UFLS), must

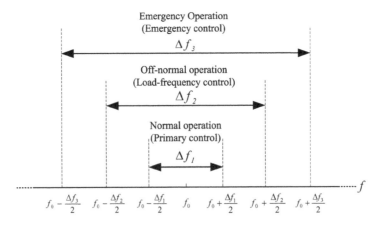

FIGURE 2.5
Frequency deviations and associated operating controls.

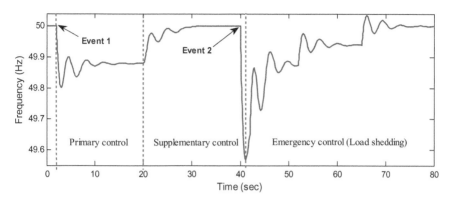

FIGURE 2.6
An example of responses of primary, supplementary, and emergency controls.

be used to decrease the risk of cascade faults, additional generation events, load/network, and separation events.

Figure 2.6 shows an example of a typical power system response to a power plant trip event, with the responses of primary, supplementary, and emergency controls. Following event 1, the primary control loops of all generating units respond within a few seconds. As soon as the balance is reestablished, the system frequency stabilizes and remains at a fixed value, but differs from the nominal frequency because of the *droop* of the generators, which provide a proportional type of action that will be explained later. Consequently, the tie-line power flows in a multiarea power system will differ from the scheduled values.

The supplementary control will take over the remaining frequency and power deviation after a few seconds, and can reestablish the nominal frequency and specified power cross-border exchanges by allocation of regulating power. Following event 1, the frequency does not fall too quickly, so there is time for the AGC system to use the regulation power and thus recover the load-generation balance. However, it does not happen following event 2, where the frequency is quickly dropped to a critical value. In this case, where the frequency exceeds the permissible limits, an emergency control plan such as UFLS may need to restore frequency and maintain system stability. Otherwise, due to critical underspeed, other generators may trip out, creating a cascade failure, which can cause widespread blackouts.

As mentioned above, following an imbalance between total generation and demand, the regulating units will then perform automatic frequency control actions, i.e., primary and supplementary control actions, and the balance between generation and demand will be reestablished. Using Union for the Coordination of Transmission of Electricity (UCTE) terminology,[5] in addition to supplementary (secondary) control, the AGC systems can perform another level of control named *tertiary control*. The tertiary control concept is close to the meaning of the *emergency control* term in the present text. This control is used to restore the secondary control reserve, manage eventual congestions, and bring back the frequency and tie-line power to their specified values if the supplementary reserve is not sufficient. These targets may be achieved by connection and tripping of power, redistributing the output from AGC participating units, and demand side (load) control.

The typical frequency control loops are represented in Figure 2.7, in a simplified scheme. In a large multiarea power system, all three forms of frequency control (primary, supplementary, and emergency) are usually available. The demand side also participates in frequency control through the action of frequency-sensitive relays that disconnect some loads at given frequency thresholds (UFLS). The demand side may also contribute to frequency control using a self-regulating effect of frequency-sensitive loads, such as induction motors. However, this type of contribution is not always taken into account in the calculation of the overall frequency control response. The following subsections summarize the characteristics of the three frequency control levels.

2.2.1 Primary Control

Depending on the type of generation, the real power delivered by a generator is controlled by the mechanical power output of a prime mover such as a steam turbine, gas turbine, hydro turbine, or diesel engine. In the case of a steam or hydro turbine, mechanical power is controlled by the opening or closing of valves regulating the input of steam or water flow into the turbine. Steam (or water) input to generators must be continuously regulated to

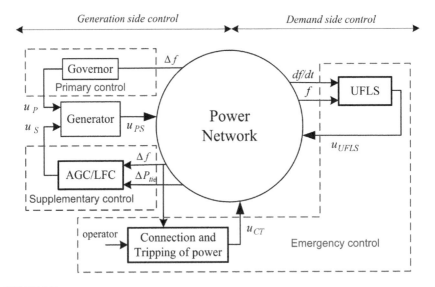

FIGURE 2.7
Frequency control loops.

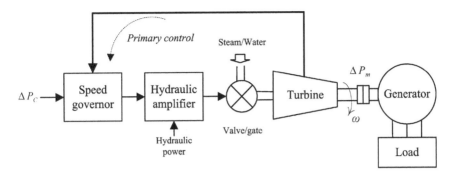

FIGURE 2.8
Governor-turbine with primary frequency control loop.

match real power demand. Without this regulation, the machine speed will vary with consequent change in frequency. For satisfactory operation of a power system, the frequency should remain nearly constant.[6]

A schematic block diagram of a synchronous generator equipped with a primary frequency control loop is shown in Figure 2.8. The *speed governor* senses the change in speed (frequency) via the primary control loop. In fact, primary control performs a local automatic control that delivers reserve power in opposition to any frequency change. The necessary mechanical forces to position the main valve against the high steam (or hydro) pressure is provided by the *hydraulic amplifier*, and the *speed changer* provides a steady-state power output setting for the turbine.[1]

The speed governor on each generating unit provides the primary speed control function, and all generating units contribute to the overall change in generation, irrespective of the location of the load change, using their speed governing. However, as mentioned, the primary control action is not usually sufficient to restore the system frequency, especially in an interconnected power system, and the supplementary control loop is required to adjust the load reference set point through the speed changer motor.

2.2.2 Supplementary Control

In addition to primary frequency control, a large synchronous generator may be equipped with a supplementary frequency control loop. A schematic block diagram of a synchronous generator equipped with primary and supplementary frequency control loops is shown in Figure 2.9.

The supplementary loop gives feedback via the frequency deviation and adds it to the primary control loop through a dynamic *controller*. The resulting signal (ΔP_C) is used to regulate the system frequency. In real-world power systems, the dynamic controller is usually a simple integral or proportional-integral (PI) controller. Following a change in load, the feedback mechanism provides an appropriate signal for the turbine to make generation (ΔP_m) track the load and restore the system frequency.

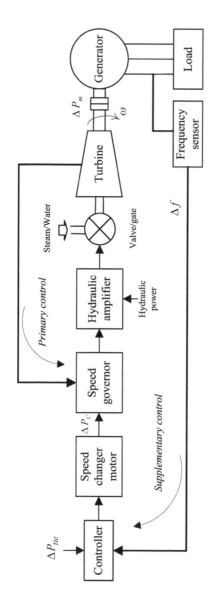

FIGURE 2.9
Frequency control mechanism.

Supplementary frequency control, which is known as load-frequency control (LFC), is a major function of AGC systems as they operate online to control system frequency and power generation. As mentioned, the AGC performance is highly dependent on how the participant generating units would respond to the control action signals. The North American Electric Reliability Council (NERC) separated generator actions into two groups. The first group is associated with large frequency deviations where generators respond through governor action and then in response to AGC signals, and the second group is associated with a continuous regulation process in response to AGC signals only. During a sudden increase in area load, the area frequency experiences a transient drop. At the transient state, there are flows of power from other areas to supply the excess load in this area. Usually, certain generating units within each area are on regulation to meet this load change. At steady state, the generation is closely matched with the load, causing tie-line power and frequency deviations to drop to zero.[7]

Several frequency control criteria and standards are available to find how well each control area must balance its aggregate generation and load. For instance, control performance standards 1 and 2 (CPS1 and CPS2) were introduced by NERC to achieve the optimum AGC performance.[8,9] CPS1 and CPS2 are measurable and can be fixed as normal functions of EMS unit in each control area. Measurements are taken continuously, with data recorded at each minute of operation. CPS1 indicates the relationship between the area control error (ACE) and the system frequency on a 1 min basis; it is the measure of short-term error between load and generation. CPS1's performance will be good if a control area closely matches generation with the load, or if the mismatch causes system frequency to be driven closer to the nominal frequency. CPS1's performance will be degraded if the system frequency is driven away from the nominal frequency.

CPS2 will place boundaries on CPS1 to limit net unscheduled power flows that are unacceptably large. Actually, it sets limits on the maximum average ACE for every 10 min period. CPS2 will prevent excessive generation/load mismatches even if a mismatch is in the proper direction. Large mismatches can cause excessive power flows and potential transmission overloads between areas with overgeneration and those with insufficient generation.[7]

2.2.3 Emergency Control

Emergency control, such as *load shedding*, shall be established in emergency conditions to minimize the risk of further uncontrolled separation, loss of generation, or system shutdown. Load shedding is an emergency control action to ensure system stability, by curtailing system load. The load shedding will only be used if the frequency (or voltage) falls below a specified frequency (voltage) threshold. Typically, the load shedding protects against excessive frequency (or voltage) decline by attempting to balance real (reactive) power supply and demand in the system.

The load shedding curtails the amount of load in the power system until the available generation can supply the remaining loads. If the power system is unable to supply its active (reactive) load demands, the under-frequency (under-voltage) condition will be intense. The number of load shedding steps, the amount of load that should be shed in each step, the delay between the stages, and the location of shed load are the important objects that should be determined in a load shedding algorithm. A load shedding scheme is usually composed of several stages. Each stage is characterized by frequency/voltage threshold, amount of load, and delay before tripping. The objective of an effective load shedding scheme is to curtail a minimum amount of load, and provide a quick, smooth, and safe transition of the system from an emergency situation to a normal equilibrium state.[1]

The interested load shedding type in power system frequency control is UFLS. Most common UFLS schemes, which involve shedding, predetermine the amounts of load if the frequency drops below specified frequency thresholds. There are various types of UFLS schemes discussed in the literature and applied by the electric utilities around the world. A classification divides the existing schemes into *static* and *dynamic* (or *fixed* and *adaptive*) load shedding types. Static load shedding curtails the constant block of load at each stage, while dynamic load shedding curtails a dynamic amount of load by taking into account the magnitude of disturbance and dynamic characteristics of the system at each stage. Although the dynamic load shedding schemes are more flexible and have several advantages, most real-world load shedding plans are of the static type.[1]

There are two basic paradigms for load shedding: a *shared* LS paradigm and a *targeted* LS paradigm. The first paradigm appears in the well-known UFLS schemes, and the second paradigm includes some recently proposed wide area LS approaches.[10] Sharing load shedding responsibilities (such as induced by UFLS) are not necessarily an undesirable feature and can be justified on a number of grounds. For example, shared load shedding schemes tend to improve the security of the interconnected regions by allowing generation reserve to be shared. Further, load shedding approaches can be indirectly used to preferentially shed the least important load in the system. However, sharing load shedding can have a significant impact on interregion power flows and, in certain situations, might increase the risk of cascade failure. Although both shared and targeted load shedding schemes may be able to stabilize the overall system frequency, the shared load shedding response leads to a situation requiring more power transmission requirements. In some situations, this increased power flow might cause line overloading and increase the risk of cascade failure.[1]

Some useful guideline for UFLS strategies can be found in IEEE.[11] The UFLS schemes typically curtail a predetermined amount of load at specific frequency thresholds. The frequency thresholds are also biased, using a disturbance magnitude to shed load at higher-frequency levels in dangerous contingencies.[10,12,13] The UCTE recommends that its members initiate the first stage of automatic UFLS in response to a frequency threshold not lower than

49 Hz.[5] Based on this recommendation, in case of a frequency drop of 49 Hz, the automatic UFLS begins with a minimum of 10 to 20% of the load. In case of lower frequency, the synchronously interconnected network may be divided into partial networks (islanding). The UFLS is performed at trigger frequencies to curtail the amount of the load (usually about 5 to 20%).

The emergency control schemes and protection plans are usually represented using incremented/decremented step behavior.[1] For instance, according to Figure 2.7, for a fixed UFLS scheme, the function of u_{UFLS} in the time domain could be considered a sum of the incremental step functions of $\Delta P_j u(t - t_j)$, as shown in Figure 2.10a. Here, ΔP_j and t_j denote the incremental amount of load shed and the time instant of the jth load shedding step, respectively. Therefore, for L load shedding steps,

$$u_{UFLS}(t) = \sum_{j=0}^{L} \Delta P_j u(t - t_j) \tag{2.1}$$

Similarly, to formulate the other emergency control schemes, such as connection and tripping of power plants, u_{CT} (Figure 2.7), appropriate step functions can be used. Therefore, using the Laplace transformation, one can represent the emergency control effect u_{EC} in the following abstracted form:

$$u_{EC}(s) = u_{UFLS}(s) + u_{CT}(s) = \sum_{l=0}^{N} \frac{\Delta P_l}{s} e^{-t_l s} \tag{2.2}$$

where ΔP_l is the size of the equivalent step load/power changes due to a generation/load event or a load shedding scheme at t_l.

As an example, to show the role of load shedding in stabilizing the power system, the dynamic behavior (voltage-frequency trajectory) of a standard nine-bus IEEE test system, following a serious disturbance (tripping of the largest generator), and applying an intelligent load shedding scheme,[13] is shown in Figure 2.10b. The load shedding steps are determined by several ellipses, and when the phase trajectory reaches each ellipse, the corresponding load shedding step is triggered. The trajectory is represented in the following complex plane:

$$S = \overset{*}{\Delta f} + j \overset{*}{\Delta v} \tag{2.3}$$

where

$$\overset{*}{\Delta f} = \frac{\Delta f}{f_0}, \ \overset{*}{\Delta v} = \frac{\Delta v}{v_0} \tag{2.4}$$

Here f_0 and v_0 are the frequency and voltage before contingency. The system frequency (and voltage) is reestablished following the five steps of load shedding.

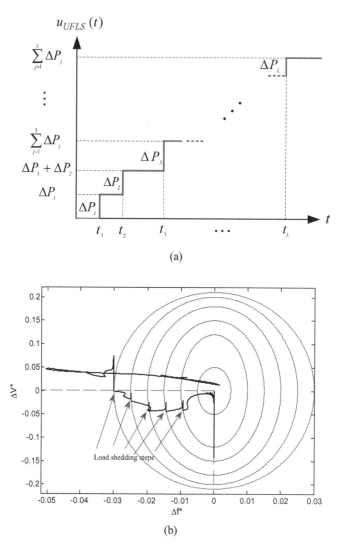

FIGURE 2.10

Load shedding: (a) L-step UFLS and (b) voltage-frequency trajectory following load shedding.

2.3 Frequency Response Model and AGC Characteristics

In an interconnected power system the *control area* concept needs to be used for the sake of synthesis and analysis of the AGC system. The control area is a coherent area consisting of a group of generators and loads, where all the generators respond to changes in load or speed changer settings, in unison.

The frequency is assumed to be the same in all points of a control area. A multiarea power system comprises areas that are interconnected by high-voltage transmission lines or tie-lines. The AGC system in an interconnected power system should control the area frequency as well as the interchange power with the other control areas.

An appropriate frequency response model for a control area *i* in a multiarea power system is shown in Figure 2.11. In AGC practice, to clear the fast changes and probable added noises, system frequency gradient and ACE signals must be filtered before being used.[1] If the ACE signal exceeds a threshold at interval T_W, it will be applied to the controller block. The controller can be activated to send higher/lower pulses to the participant generation units if its input ACE signal exceeds a standard limits. Delays, ramping rate, and range limits are different for various generation units. Concerning the limit on generation, governor dead-band, and time delays, the AGC model becomes highly nonlinear; hence, it will be difficult to use the conventional linear techniques for performance optimization and control design.

2.3.1 Droop Characteristic

The ratio of speed (frequency) change (Δf) to change in output-generated power (ΔP_g) is known as *droop* or speed regulation, and can be expressed as

$$R\left(Hz\!\!\Big/\!\!pu.MW\right) = \frac{\Delta f}{\Delta P_g} \tag{2.5}$$

For example, a 5% droop means that a 5% deviation in nominal frequency (from 60 to 57 Hz) causes a 100% change in output power. In Figure 2.11, the droop characteristics for the generating units (R_{ki}) are properly shown in the primary frequency control loops.

The interconnected generating units with different droop characteristics can jointly track the load change to restore the nominal system frequency. This is illustrated in Figure 2.12, representing two units with different droop characteristics connected to a common load. Two generating units are operating at a unique nominal frequency with different output powers. The change in the network load causes the units to decrease their speed, and the governors increase the outputs until they reach a new common operating frequency. As expressed in Equation 2.6, the amount of produced power by each generating unit to compensate the network load change depends on the unit's droop characteristic.[14,15]

$$\Delta P_{gi} = \frac{\Delta f}{R_i} \tag{2.6}$$

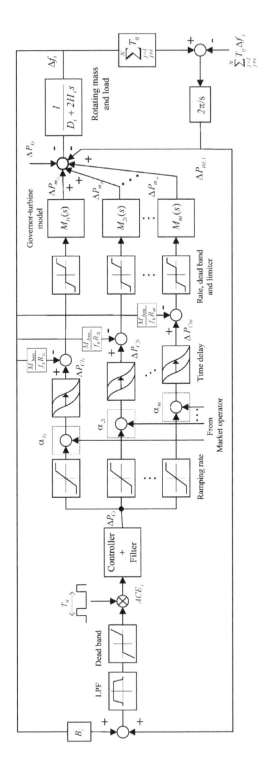

FIGURE 2.11
A frequency response model for dynamic performance analysis.

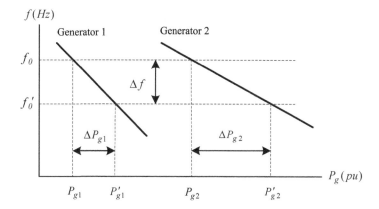

FIGURE 2.12
Load tracking by generators with different droops.

Hence,

$$\frac{\Delta P_{g1}}{\Delta P_{g2}} = \frac{R_2}{R_1} \qquad (2.7)$$

2.3.2 Generation-Load Model

For the purposes of AGC synthesis and analysis in the presence of load disturbances, a simple, low-order linearized model is commonly used. The overall generation-load dynamic relationship between the incremental mismatch power ($\Delta P_m - \Delta P_L$) and the frequency deviation (Δf) can be expressed as[1,16]

$$\Delta P_m(t) - \Delta P_L(t) = 2H\frac{d\Delta f(t)}{dt} + D\Delta f(t) \qquad (2.8)$$

where ΔP_m is the mechanical power change, ΔP_L is the load change, H is the inertia constant, and D is the load damping coefficient.

Using the Laplace transform, Equation 2.8 can be written as

$$\Delta P_m(s) - \Delta P_L(s) = 2Hs\Delta f(s) + D\Delta f(s) \qquad (2.9)$$

Equation 2.9 is represented in the right-hand side of the frequency response model described in Figure 2.11.

2.3.3 Area Interface

In a multiarea power system, the trend of frequency measured in each control area is an indicator of the trend of the mismatch power in the interconnection and not in the control area alone. Therefore, the power interchange should be properly considered in the LFC model. It is easy to show that in an interconnected power system with N control areas, the tie-line power change between area i and other areas can be represented as[1]

$$\Delta P_{tie,i} = \sum_{\substack{j=1 \\ j \neq i}}^{N} \Delta P_{tie,ij} = \frac{2\pi}{s} \left[\sum_{\substack{j=1 \\ j \neq i}}^{N} T_{ij}\Delta f_i - \sum_{\substack{j=1 \\ j \neq i}}^{N} T_{ij}\Delta f_j \right] \qquad (2.10)$$

where $\Delta P_{tie,i}$ indicates the tie-line power change of area i, and T_{12} is the synchronizing torque coefficient between areas i and j. Equation 2.10 is also realized in the bottom-right side of the AGC block diagram in Figure 2.11. The effect of changing the tie-line power for an area is equivalent to changing the load of that area. That is why in Figure 2.11, the $\Delta P_{tie,i}$ has been added to the mechanical power change (ΔP_m) and area load change (ΔP_L) using an appropriate sign.

In addition to the regulating area frequency, the LFC loop should control the net interchange power with neighboring areas at scheduled values. This is generally accomplished by feeding a linear combination of tie-line flow and frequency deviations, known as *area control error* (ACE), via supplementary feedback to the dynamic controller. The ACE can be calculated as follows:

$$ACE_i = \Delta P_{tie,i} + \beta_i \Delta f_i \qquad (2.11)$$

where β_i is a bias factor, and its suitable value can be computed as[16]

$$\beta_i = \frac{1}{R_i} + D_i \qquad (2.12)$$

The block diagram shown in Figure 2.11 illustrates how Equation 2.11 is implemented in the supplementary frequency control loop. The effects of local load changes and interface with other areas are also considered as the following two input signals:

$$w_1 = \Delta P_{Li}, \quad w_2 = \sum_{\substack{j=1 \\ j \neq i}}^{N} T_{ij}\Delta f_j \qquad (2.13)$$

Each control area monitors its own tie-line power flow and frequency at the area control center, and the combined signal (ACE) is allocated to the dynamic controller. Finally, the resulting control action signal is applied to the turbine-governor units, according their participation factors. In Figure 2.11, $M_{ki}(s)$ and α_{ki} are the governor-turbine model and AGC participation factor for generator unit k, respectively.

2.3.4 Spinning Reserve

There are different definitions for the *spinning reserve* term. Using UCTE terminology, it is a tertiary reserve that can be available within 15 min and is provided chiefly by storage stations, pumped-storage stations, gas turbines, and thermal power stations operating at less than full output. While based on the definition provided by NERC, it is an unloaded generation that is synchronized and ready to serve additional demand.[17]

The spinning reserve can be simply defined as the difference between capacity and existing generation level. It refers to spare power capacity to provide the necessary regulation power for the sum of primary and secondary control issues. Regulation power is required power to bring the system frequency back to its nominal value. The frequency-dependent reserves are automatically activated by the AGC system, when the frequency is at a lower level than the nominal value (50 or 60 Hz, depending on the system).

Always, the market operator needs to ensure that there is enough reserved capacity for potential future occurrences. The size of the AGC reserve that is required depends on the size of load variation, schedule changes, and generating units. In a deregulated environment, the reserve levels may be influenced by the market operation. If too much energy is traded, the market operator must contract more reserves to ensure that the predicted demand can be met.[18] Additional reserves need to be activated to restore the used power spinning reserves in preparation for further incidents.

2.3.5 Participation Factor

The *participation factor* indicates the amount of participation of a generator unit in the AGC system. Following a load disturbance within the control area, the produced appropriate supplementary control signal is distributed among generator units in proportion to their participation, to make generation follow the load. In a given control area, the sum of participation factors is equal to 1:

$$\sum_{k=1}^{n} \alpha_{ki} = 1 , \quad 0 \le \alpha_{ki} \le 1 \tag{2.14}$$

In a competitive environment, AGC participation factors are actually time-dependent variables and must be computed dynamically by an independent organization based on bid prices, availability, congestion problems, costs, and other related issues.[1]

2.3.6 Generation Rate Constraint

Although considering all dynamics to achieve an accurate perception of the AGC subject may be difficult and not useful, considering the main inherent requirement and the basic constraints imposed by the physical system dynamics to model/evaluate the AGC performance is important. An important physical constraint is the rate of change of power generation due to the limitation of thermal and mechanical movements, which is known as *generation rate constraint* (GRC).[1]

Rapidly varying components of system signals are almost unobservable due to various filters involved in the process, and an appropriate AGC scheme must be able to maintain sufficient levels of reserved control range and control rate. Therefore, the rate of change in the power output of generating units used for AGC must in total be sufficient for the AGC purpose. It is defined as a percentage of the rated output of the control generator per unit of time. The generation rates for generation units, depending on their technology and types, are different. Typical ramp rates for different kinds of units (as a percentage of capacity) for diesel engines, industrial GT, GT combined cycle, steam turbine plants, and nuclear plants are 40%/min, 20%/min, 5 to 10%/min, 1 to 5%/min, and 1 to 5%/min, respectively.[19] In hard-coal-fired and lignite-fired power plants, this rate is 2 to 4%/min and 1 to 2%/min, respectively.[5]

2.3.7 Speed Governor Dead-Band

If the input signal of a speed governor is changed, it may not immediately react until the input reaches a specified value. This phenomenon is known as *speed governor dead-band*. All governors have a dead-band in response, which is important for AGC systems. Governor dead-band is defined as the total magnitude of a sustained speed change, within which there is no resulting change in valve position.

The maximum value of dead-band for governors of large steam turbines is specified as 0.06% (0.036 Hz).[20] For a wide dead-band the AGC performance may be significantly degraded. An effect of the governor dead-band on the AGC operation is to increase the apparent steady-state frequency regulation. In Figure 2.11, the GRC and speed governor dead-band are considered by adding limiters and hysteresis patterns to the governor-turbine system models.

2.3.8 Time Delays

In new power systems, *communication delays* are becoming a more significant challenge in system operation and control. Although, under a traditional AGC structure, the problems associated with the communication links may ignorable, considering the problems that may arise in the communication system in use of an open communication infrastructure to support the ancillary services in a restructured environment is important. It has been shown that time delays can degrade the AGC performance seriously.[21] The AGC performance declines when the time delay increases.

The time delays in the AGC systems mainly exist on the communication channels between the control center and operating stations—specifically on the measured frequency and power tie-line flow from RTUs or IEDs to the control center, and the delay on the produced rise/lower signal from the control center to individual generation units.[22] Furthermore, all other probable data communication, signal processing, and filtering among an AGC system introduce delays that should be considered. These delays are schematically shown in Figure 2.11.

2.4 A Three-Control Area Power System Example

To illustrate the system frequency response in a multiarea power system based on the model described in Figure 2.11, consider three identical interconnected control areas, as shown in Figure 2.13. The simulation parameters are given in Table 2.1. Here, the Mega-Volt-Ampere (MVA) base is 1,000, and each control area uses a PI controller in its supplementary frequency control loop.

The system response following a simultaneous 0.05 pu load step (disturbance) increase at 2 s in control areas 1 and 2 is shown in Figures 2.14 to 2.17. Although the load disturbances occur in areas 1 and 2, area 3 also participates in restoring the system frequency and minimizing the tie-line power fluctuation using generating units G_8 and G_9.

Several low-order models for representing turbine-governor dynamics, $M_i(s)$, for use in power system frequency analysis and control design are

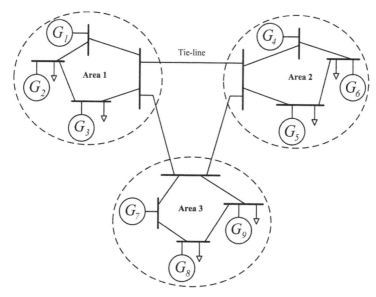

FIGURE 2.13
Three-control area power system.

TABLE 2.1

Simulation Parameters for Three-Control Area Power System

Parameters Generating Units	Area 1			Area 2			Area 3		
	G_1	G_2	G_3	G_4	G_5	G_6	G_7	G_8	G_9
Rate (MW)	1,000	800	1,000	1,100	900	1,200	850	1,000	1,020
T_{tk} (s)	0.4	0.36	0.42	0.44	0.32	0.40	0.30	0.40	0.41
T_{gk} (s)	0.08	0.06	0.07	0.06	0.06	0.08	0.07	0.07	0.08
Participation factor α_i	0.4	0.4	0.2	0.6	0	0.4	0	0.5	0.5
Ramp rate (MW/min)	8	8	6	12	10	8	10	10	10
Dead zone band	0.02	0.02	0.02	0.02	0.02	0.02	0.02	0.02	0.02
Saturation limits (pu)	±0.1	±0.1	±0.1	±0.1	±0.1	±0.1	±0.1	±0.1	±0.1
Time delay (s)	1			1			1		
B_i (pu/Hz)	1.0136			1.1857			1.0735		
D_i (pu MW/Hz)	0.044			0.044			0.046		
R_i (Hz/pu)	3.00			2.67			2.95		
H_i	0.2433			0.2739			0.2392		
Controller	$-0.01 - (0.19/s)$			$-0.03 - (0.25/s)$			$-0.02 - (0.27/s)$		

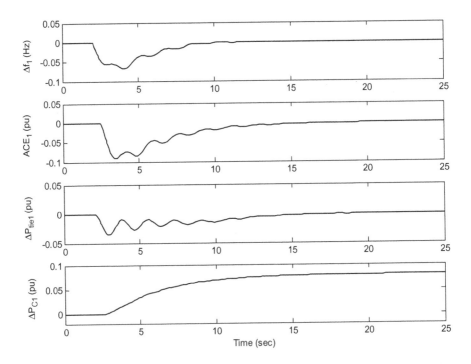

FIGURE 2.14

The system response in control area 1.

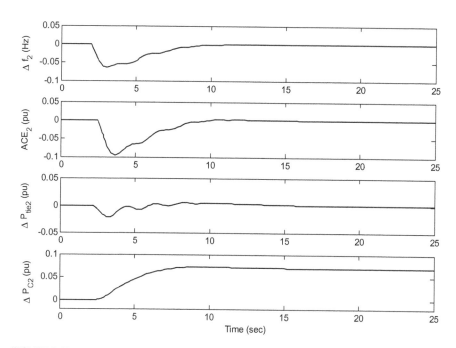

FIGURE 2.15
The system response in control area 2.

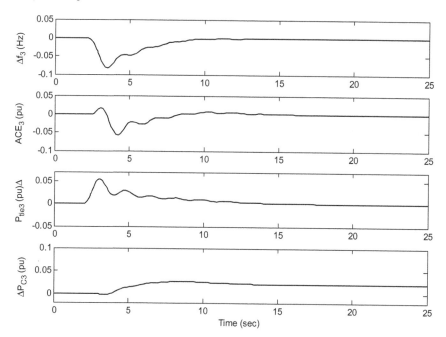

FIGURE 2.16
The system response in control area 3.

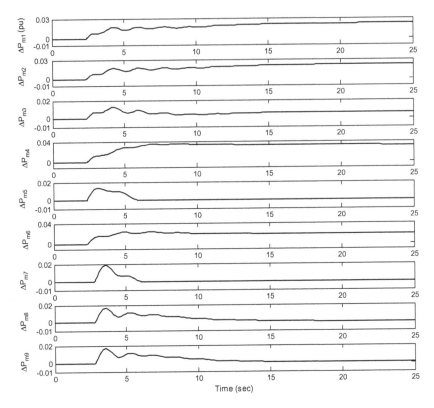

FIGURE 2.17
Mechanical power changes in the generating units.

introduced in Bevrani.[1] For the present example, it is assumed that all generators are nonreheat steam units; therefore, the turbine-governor dynamics can be approximated by[1]

$$M_{ki}(s) = \frac{1}{(1+T_{gk}s)} \cdot \frac{1}{(1+T_{tk}s)}$$ (2.15)

where T_{gk} and T_{tk} are governor and turbine time constants, respectively. The balance between connected control areas is achieved by detecting the frequency and tie-line power deviations to generate the ACE signal, which is in turn utilized in a dynamic controller.

The frequency response model, which is described in Figure 2.11, is implemented for each control area in the MATLAB software. Figures 2.14 to 2.16 show the frequency deviation, ACE, tie-line power change, and control action signal for control areas 1 to 3, respectively. The proposed simulation shows the supplementary frequency control loops properly act to maintain system frequency and exchange powers close to the scheduled values by sending a corrective signal to the generating units in proportion to their participation in the AGC system.

The difference between the starting times in simulations is because of considering a small communication delay (about 1 s). This delay is needed for producing the ACE and the control action signals in the control center following a disturbance. Figure 2.17 shows the mechanical power fluctuation in all generating units following the simultaneous 0.05 step load disturbance in areas 1 and 2.

Figure 2.17 indicates that the mechanical power to compensate the frequency deviation and tie-line power change initially comes from all generating units to respond to the step load increase in areas 1 and 2, and results in a frequency drop sensed by the speed governors of all generators. However, after a few seconds (at steady state), the additional powers against the local load changes come only from generating units that are participating in the AGC issue.

The amount of additional generated power by each unit is proportional to the related participation factor. Figure 2.17 shows that the participation factors for generating units G_5 and G_7 is zero, while the maximum participation belongs to generating unit G_4. These results agree with the data given in Table 2.1.

2.5 Summary

The AGC issue, with definitions provided in cooperation with SCADA and EMS, and basic concepts are addressed. The AGC mechanism in an interconnected power system is described. The important AGC characteristics and physical constraints are explained. The impacts of generation rate, deadband, and time delays on the AGC performance are emphasized. Finally, a suitable dynamic frequency response model is introduced, and in order to understand the dynamic behavior of AGC in a multiarea power system, some simulations are performed.

References

1. H. Bevrani. 2009. *Robust power system frequency control*. New York: Springer.
2. N. K. Stanton, J. C. Giri, A. Bose. 2007. Energy management. In *Power system stability and control*, ed. L. L. Grigsby. Boca Raton, FL: CRC Press.
3. IEEE PES. 2008. *Standard for SCADA and automation systems*. IEEE Standard C37.1.

4. H. Bindner, O. Gehrke. 2009. System control and communication. In *RisØ enery report: The intelligent energy system infrastructure for the future*, ed. H. Larsen, L. S. Petersen, 39–42. Vol. 8. National Laboratory for Sustainable Energy, Roskilde, Denmark.

5. *UCTE operation handbook*. 2009. http://www.ucte.org.

6. T. K. Nagsarkar, M. S. Sukhija. 2007. *Power system analysis*. New Delhi, IN: Oxford University Press.

7. M. Shahidehpour, H. Yamin, Z. Li. 2002. *Market operations in electric power systems: Forecasting, scheduling, and risk management*. New York: John Wiley & Sons.

8. NERC. 2002. NERC operating manual. Princeton, NJ.

9. N. Jaleeli, L. S. Vanslyck. 1999. NERC's new control performance standards. *IEEE Trans. Power Syst.* 14(3):1092–99.

10. J. J. Ford, H. Bevrani, G. Ledwich. 2009. Adaptive load shedding and regional protection. *Int. J. Electrical Power Energy Syst.* 31:611–18.

11. IEEE. 2007. Guide *for the application of protective relays used for abnormal frequency load shedding and restoration*, c1–43. IEEE Standard C37.117.

12. H. Bevrani, G. Ledwich, Z. Y. Dong, J. J. Ford. 2009. Regional frequency response analysis under normal and emergency conditions. *Electric Power Syst. Res.* 79:837–45.

13. H. Bevrani, A. G. Tikdari. 2010. An ANN-based power system emergency control scheme in the presence of high wind power penetration. In *Wind power systems: Applications of computational intelligence*, ed. L. F. Wang, C. Singh, A. Kusiak, 215–54. Heidelberg: Springer-Verlag.

14. A. J. Wood, B. F. Wollenberg. 1996. *Power generation, operation and control*. 2nd ed. New York: John Wiley & Sons.

15. D. Das. 2006. *Electric power systems*. New Delhi, IN: New Age International Ltd.

16. P. Kundur. 1994. *Power system stability and control*. New York: McGraw-Hill.

17. Y. Rebours. 2008. A comprehensive assessment of markets for frequency and voltage control ancillary services. PhD dissertation, University of Manchester.

18. G. A. Chown, B. Wigdorowitz. 2004. A methodology for the redesign of frequency control for AC networks. *IEEE Trans. Power Syst.* 19(3):1546–54.

19. Power Systems Engineering Research Center (PSERC). 2009. *Impact of increased DFIG wind penetration on power systems and markets*. Final project report.

20. IEEE. 1992. *Recommended practice for functional and performance characteristics of control systems for steam turbine-generator units*. IEEE Standard 122-1991.

21. H. Bevrani, T. Hiyama. 2009. On robust load-frequency regulation with time delays: Design and real-time implementation. *IEEE Trans. Energy Conversion* 24(1):292–300.

22. H. Bevrani, T. Hiyama. 2007. Robust load-frequency regulation: A real-time laboratory experiment. *Optimal Control Appl. Methods* 28(6):419–33.

3

Intelligent AGC: Past Achievements and New Perspectives

Automatic generation control (AGC) synthesis and analysis in power systems has a long history and its literature is voluminous. The preliminary AGC schemes have evolved over the past decades, and interest continues in proposing new intelligent AGC approaches with an improved ability to maintain tie-line power flow and system frequency close to specified values.

The first attempts in the area of AGC are given in several references.[1-4] Then the standard definitions of the terms associated with power systems AGC were provided by the IEEE working group.[5] The first optimal control concept for megawatt-frequency control design of interconnected power systems was addressed by Elgerd and Fosha.[6,7] According to physical constraints and to cope with the changed system environment, suggestions for dynamic modeling and modifications to the AGC definitions were given from time to time over the past years.[8,9] System nonlinearities and dynamic behaviors such as governor dead-band and generation rate constraint, load characteristics, and the interaction between the frequency (real power) and voltage (reactive power) control loops for the AGC design procedure have been considered.[10-14] The AGC analysis/modeling, special applications, constraints formulation, frequency bias estimation, model identification, and performance standards have led to the publishing of numerous reports.[15-24]

The AGC analysis and synthesis has been augmented with valuable research contributions during the last few decades. Significant improvements have appeared in the area of AGC designs to cope with uncertainties, various load characteristics, changing structure, and integration of new systems, such as energy storage devices, wind turbines, photovoltaic cells, and other sources of electrical energy.[25] Numerous analog and digital control schemes using nonlinear and linear optimal/robust, adaptive, and intelligent control techniques have been presented. The most recent advance in the AGC synthesis to tackle the difficulty of using complex/nonlinear power system models or insufficient knowledge about the system is the application of intelligent concepts such as neural networks, fuzzy logic, genetic algorithms, multiagent systems, and evolutionary and heuristic optimization techniques. A survey and exhaustive bibliography on the AGC have been published.[25-27]

Since Elgerd and Fosha's work,[6,7] extensive research has been done on the application of modern control theory to design more effective supplementary controllers. Several AGC synthesis approaches using optimal control techniques have been presented.[28–35] The efforts were usually directed toward the application of suitable linear state feedback controllers to the AGC problem. They have mainly optimized a constructed cost function to meet AGC objectives by well-known optimization techniques. Since an optimal AGC scheme needs the availability of all state variables, some developed strategies have used state estimation using an observer. Due to technical constraints in the design of AGC using all state variables, suboptimal AGC systems were introduced. Apart from optimal/suboptimal control strategies, the concept of variable structure systems has also been used to design AGC regulators for power systems.[36–39] These approaches enhance the insensitivity of an AGC system to parameter variations.

Since parametric uncertainty is an important issue in AGC design, the application of robust control theory to the AGC problem in multiarea power systems has been extensively studied during the last two decades.[25,40–43] The main goal is to maintain robust stability and robust performance against system uncertainties and disturbances. For this purpose, various robust control techniques such as H∞, linear matrix inequalities (LMIs), Riccati equation approaches, Kharitonov's theorem, structured singular value (μ) theory, quantitative feedback theory, Lyapunov stability theory, pole placement technique, and Q-parameterization have been used.

Apart from these design methodologies, adaptive and self-tuning control techniques have been widely used for power system AGC design during the last three decades.[44–46] The major part of the work reported so far has been performed by considering continuous time power system models. The digital and discrete type frequency regulator is also reported in some work.[13,32,46–48] Few publications have appeared on the application (or in the presence) of special devices such as superconductivity magnetic energy storage (SMES) and solid-state phase shifter.[49] The increasing need for electrical energy, limited fossil fuel reserves, and increasing concerns to environmental issues call for a fast development in the area of renewable energy sources (RESs). Some recent studies analyze the impacts of battery energy storage (BES), photovoltaic (PV) power generation, capacitive energy, and wind turbines on the performance of the AGC system, or their application in power system frequency control.[25] Considerable research on the AGC incorporating a high voltage direct current (HVDC) link is contained in Yoshida and Machida[50] and Sanpei et al.[51]

The intelligent technology offers many benefits in the area of complex and nonlinear control problems, particularly when the system is operating over an uncertain operating range. Generally, for the sake of control synthesis, nonlinear systems such as power systems are approximated by reduced order dynamic models, possibly linear, that represent the simplified dominant system's characteristics. However, these models are only valid within specific operating ranges, and a different model may be required in the case of changing operating conditions. On the other hand, due to increasing the size and complexity of modern power systems, classical and nonflexible AGC structures may not represent desirable performance over a wide range of operating conditions. Therefore, more flexible and intelligent approaches are needed.

In recent years, following the advent of modern intelligent methods, such as artificial neural networks (ANNs), fuzzy logic, multiagent systems, genetic algorithms (GAs), expert systems, simulated annealing (SA), Tabu search, particle swarm optimization, ant colony optimization, and hybrid intelligent techniques, some new potentials and powerful solutions for AGC synthesis have arisen. The human and nature ability to control complex organisms has encouraged researchers to pattern controls on human/nature responses, fuzzy behaviors, and neural network systems. Since all of the developed artificial intelligent techniques are usually dependent on knowledge extracted from environment and available data, *knowledge management* plays a pivotal role in the AGC synthesis procedures.

In the present chapter, the application of intelligent techniques on AGC synthesis is emphasized, and basic control configurations with recent achievements are briefly discussed. New challenges and key issues concerning system restructuring and integration of distributed generators and renewable energy sources are also discussed. The chapter is organized as follows: The applications of fuzzy logic, neural networks, genetic algorithms, multiagent systems, and combined intelligent techniques and evolutionary optimization approaches on the AGC synthesis problem are reviewed in Sections 3.1 to 3.5, respectively. An introduction to AGC design in deregulated environments is given in Section 3.6. AGC analysis and synthesis in the presence of renewable energy sources (RESs) and microgrids, including literature review, present worldwide status, impacts, and technical challenges, are presented in Sections 3.7 and 3.8. Finally, a discussion on the future works and research needs is given in Section 3.9.

3.1 Fuzzy Logic AGC

Nowadays, because of simplicity, robustness, and reliability, fuzzy logic is used in almost all fields of science and technology, including solving a wide range of control problems in power system control and operation. Unlike the traditional control theorems, which are essentially based on the linearized mathematical models of the controlled systems, the fuzzy control methodology tries to establish the controller directly based on the measurements, long-term experiences, and knowledge of domain experts/operators.

Several studies have been reported for the fuzzy-logic-based AGC design schemes in the literature.[52–70] There are many possible fuzzy logic controller structures for AGC purposes, some differing significantly from each other by the number and type of inputs and outputs, or less significantly by the number and type of input and output fuzzy sets and their membership functions, or by the type of control rules, inference engine, and defuzzification method. In fact, it is up to the designer to decide which controller structure would be optimal for the AGC problem. The applications of fuzzy logic in AGC systems can be classified into three categories: (1) using a fuzzy logic system as a dynamic fuzzy controller called fuzzy logic controller (FLC),[55,58,61,64–66,68] (2) using a fuzzy logic system in the form of a proportional integral (PI) (or proportional-integral-derivative [PID]) controller, or as a primer for tuning the gains of the existing PI (or PID) controller,[53,55,57,59,62,69] and (3) using fuzzy logic for other AGC aspects, such as economic dispatching.[70] Next, the first two categories are briefly described.

3.1.1 Fuzzy Logic Controller

A general scheme for a fuzzy-logic-based AGC system is given in Figure 3.1. As shown, the fuzzy controller has four blocks. Crisp input information (usually measured by area control error (ACE) or frequency deviation) from the control area is converted into fuzzy values for each input fuzzy set with the fuzzification block. The universe of discourse of the input variables determines the required scaling/normalizing for correct per-unit operation. The inference mechanism determines how the fuzzy logic operations are performed and, together with the knowledge base, the outputs of each fuzzy if-then rule. Those are combined and converted to crispy values with the defuzzification block. The output crisp value can be calculated by the center of gravity or the weighted average; then the scaled output as a control signal is applied to the generating units.

Generally, a controller design based on fuzzy logic for a dynamical system involves the following four steps:

1. Understanding of the system dynamic behavior and characteristics. Define the states and input/output control variables and their variation ranges.
2. Identify appropriate fuzzy sets and membership functions. Create the degree of fuzzy membership function for each input/output variable and complete fuzzification.
3. Define a suitable inference engine. Construct the fuzzy rule base, using the control rules that the system will operate under. Decide how the action will be executed by assigning strengths to the rules.
4. Determine defuzzification method. Combine the rules and defuzzify the output.

Consistent with the AGC design, the first step of fuzzy controller design is to choose the correct input signals to the AGC. The ACE and its derivative are usually chosen as inputs of the fuzzy controller. These two signals are then used as rule-antecedent (if-part) in the formation of rule base, and the control output is used to represent the contents of the rule-consequent (then-part) in performing the rule base.

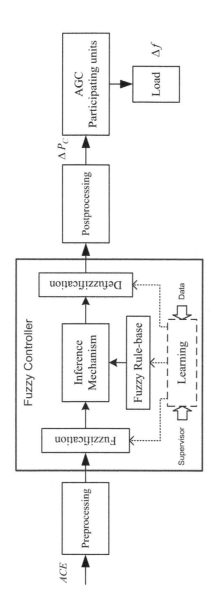

FIGURE 3.1
A general scheme for fuzzy-logic-based AGC.

Normalization or scale transformation, which maps the physical values of the current system state variables into a normalized universe of discourse, should be properly considered. This action is also needed to map the normalized value of control output variables into its physical domain (denormalization output). The normalization can be obtained by dividing each crisp input on the upper boundary value for the associated universe.

In the real world, many phenomena and measures are not crisp and deterministic. *Fuzzification* plays an important role in dealing with uncertain information, which might be objective or subjective in nature. The fuzzification block in the fuzzy controller represents the process of making a crisp quantity fuzzy. In fact, the fuzzifier converts the crisp input to a linguistic variable using the membership functions stored in the fuzzy knowledge base. Fuzzines in a fuzzy set are characterized by the *membership functions*. Using suitable membership functions, the ranges of input and output variables are assigned linguistic variables. These variables transform the numerical values of the input of the fuzzy controller to fuzzy quantities. These linguistic variables specify the quality of the control. Triangular, trapezoid, and Gaussian are the more common membership functions used in fuzzy control systems.

A *knowledge rule base* consists of information storage for linguistic variable definitions (database) and fuzzy rules (control base). The concepts associated with a database are used to characterize fuzzy control rules and a fuzzy data manipulation in a fuzzy logic controller. A lookup table based on discrete universes defines the output of a controller for all possible combinations of the input signals. A fuzzy system is characterized by a set of linguistic statements in the form of if-then rules. Fuzzy conditional statements make the rules or the rule set of the fuzzy controller. Finally, the *inference engine* uses the if-then rules to convert the fuzzy input to the fuzzy output.

On the other hand, a defuzzifier converts the fuzzy output of the inference engine to crisp values using membership functions analogous to the ones used by the fuzzifier. For the *defuzzification* process, the center of sums, mean-max, weighted average, and centroid methods are commonly employed to defuzzify the fuzzy incremental control law. The parameters of the fuzzy logic controller, such as membership functions, can be adjusted using an external tuning mechanism. The resulting controller is known as an adaptive, self-learning, or self-tuning fuzzy controller. An adaptive fuzzy controller has a distinct architecture consisting of two loops: an inner control loop, which is the basic feedback loop, and an outer loop, which adjusts the parameters of the controller. This architecture is shown in Figure 3.2.

The adaptive fuzzy controllers commonly use some other intelligent techniques, such as neural networks, which have a learning capability. In this case, new control configurations such as neuro-fuzzy control appear. A tuning mechanism may use a reference model to provide the tuning/learning signal for changing the core of the fuzzy controller. In this way, the self-organization process enforces the fuzzy control system to follow the given reference model dynamics.

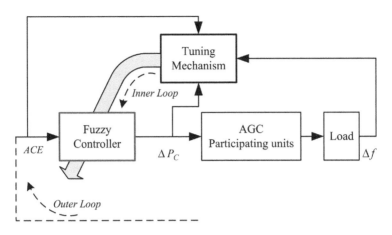

FIGURE 3.2
A general scheme for adaptive fuzzy logic AGC.

3.1.2 Fuzzy-Based PI (PID) Controller

The PI and PID control structures have been widely used in power system control due to their design/structure simplicity and inexpensive cost. Among these controllers, the most commonly used one in AGC systems is the PI. The success of the PI controller depends on an appropriate choice of its gains. The control signal generated by a PI controller in the continuous time domain is represented by

$$u(t) = k_P ACE(t) + k_I \int_0^t ACE(\tau)d\tau \tag{3.1}$$

where $u(t) = \Delta P_C(t)$ is the control signal, $ACE(t)$ is the (area control) error signal, and k_P and k_I are the proportional and integral coefficients, respectively. For ease of digital implementation, a PI controller in the discrete time domain can be described by one of the following forms:

$$u(kT) = k_P ACE(kT) + k_I \left\{ T \sum_{i=1}^n ACE(iT) \right\} \tag{3.2}$$

$$\Delta u(kT) = u(kT) - u\left[(k-1)T\right] = k_P \left[\frac{ACE(kT) - ACE\left[(k-1)T\right]}{T} \right] + k_I ACE(kT)$$

$$\tag{3.3}$$

where T is the sampling period. The second term in Equation 3.2 is the forward rectangular integration approximation of the integral term in Equation 3.1.

In AGC practice, tuning the PI gains is usually realized by experienced human experts; therefore, it may not be possible to achieve a desirable performance for AGC in large-scale power systems with high order, time delays, nonlinearities, and uncertainties, and without precise mathematical models. The fuzzy PI controllers are introduced to improve the performance of the AGC systems in comparison to conventional PI tuning methods. It has been found that the AGC systems with fuzzy-logic-based PI controllers have better capabilities of handling the aforesaid systems.

Next, the components of a fuzzy PI controller in an AGC system are briefly discussed. The fuzzy PI controller based on Equation 3.3 is shown in Figure 3.3. N_P, N_I, and $N_{\Delta u}$ are the normalization factors for ACE_p, ACE_i, and Δu, respectively. $N_{\Delta u}^{-1}$ is the reciprocal of $N_{\Delta u}$, called a denormalization factor. These factors play a role similar to that of the gain coefficients k_p and k_I in a conventional PI controller. The fuzzy PI controller employs two inputs: the error signal $ACE(kT)$, and the first-order time derivative of $ACE(kT)$. It has a single control output $u(t) = \Delta P_C(t)$; N_P and N_I are the normalized inputs, and $N_{\Delta u}$ is the normalized output.

It is noteworthy that the PI (PID) fuzzy controllers require a two- (three-) dimensional rule base. This issue makes the AGC design process more difficult. To reduce the number of interactive fuzzy relations among subsystems, the concept of decomposition of multivariable systems for the purpose of distributed fuzzy control design can be used.[61]

In addition to the above fuzzy PI/PID-based AGC design, some works use fuzzy logic to tune the gain parameters of existing PI/PID controllers in the AGC systems.[53,55] The mentioned control scheme is conceptually shown for a PI controller in Figure 3.4. The details of fuzzy-logic-based AGC design with the implementation process are presented in Chapter 9.

3.2 Neuro-Fuzzy and Neural-Networks-Based AGC

As described in Figure 3.2, to perform a self-tuning/adaptive fuzzy controller, one may use the learning capability of neural networks in the block of a tuning mechanism. This combination provides a type of *neuro-fuzzy* controller. A neuro-fuzzy controller is a fuzzy controller that uses a *learning algorithm* inspired by the *neural network* theory to determine its *parameters* by processing data samples. The combined intelligent-based AGC design using ANN and fuzzy logic techniques is presented in several works to utilize the novel aspects of both designs in a single hybrid AGC system.[63,71]

Neural networks are numerical model-free estimators, which can estimate from sample data how output functionally depends on input without the need for complex mathematical models. An ANN consists of a number of nonlinear computational processing elements (neurons) arranged in

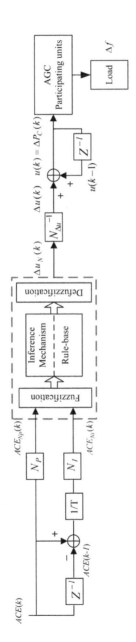

FIGURE 3.3
Fuzzy PI control scheme.

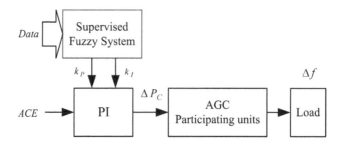

FIGURE 3.4
Fuzzy logic for tuning of PI-based AGC system.

several layers, including an input layer, an output layer, and one or more hidden layers in between. Every layer contains one or more neurons, and the output of each neuron is usually fed into all or most of the inputs of the neurons in the next layer. The input layer receives input signals, which are then transformed and propagated simultaneously through the network, layer by layer.

A neuron accepts one or more input signals and produces one output, which is a nonlinear function of the weighted sum of inputs. The mapping from the input variables to the output variables can be fixed by setting all the weights associated with each neuron to some constant. In fact, the training of an ANN in a control structure is a procedure to adjust these values so that the ANN can map all the input control values to the corresponding output control values.

The ANNs with their massive parallelism and ability to learn any type of nonlinearities are also purely used for the design of AGC systems.[14,72–80] A multilayer nonlinear network for supplementary control design using a backpropagation training algorithm is addressed in Beaufays et al.[72] The ANN control performance is compared with classical PI control design on single-area and two-area power systems. An AGC scheme to incorporate the nonconforming load problem showing an effort to develop algorithms capable of discriminating between noncontrollable short-term and long-term excursions is presented in Douglas et al.[14] The ANN techniques are used for recognition of controllable signals in the presence of a noisy random load.

The AGC system performance is evaluated with a nonlinear neural network controller using a generalized neural structure in Chaturvedi et al.[73] An ANN-based AGC system is examined on a four-control area power system considering the reheat nonlinearity effect of the steam turbine and upper/lower constraints for generation rate constraint of the hydro turbine in Zeynelgil et al.[74]

From the control configuration point of view, the most proposed ANN-based AGC designs can be divided into three general control structures that are conceptually shown in Figure 3.5: (1) using the ANN system as a controller to provide control command in the main feedback loop, (2) using ANN for tuning the parameters of an existing fixed structure controller

(I, PI, PID, etc.), and finally, (3) using the ANN system as an additional controller in parallel with the existing conventional simple controller, such as I/PI, to improve the closed-loop performance.

The aforementioned three configurations are presented in Figure 3.5a–c, respectively. *Backpropagation*, which is a gradient descent learning algorithm, is one of the most popular supervised learning algorithms in all mentioned configurations. It backpropagates the error signals from the output layer to all the hidden layers, such that their weights can be adjusted accordingly. Backpropagation is a generalization of the least mean squares (LMS) procedure for feedforward, multilayered networks with hidden layers. It uses a gradient descent technique, which changes the weights between neurons in its original and simplest form by an amount proportional to the partial derivative of the error function with respect to the given weight.

In Figure 3.5b, the ANN performs an automatic tuner. The initial values for the parameters of the fixed structure controller (e.g., k_p, k_1, and k_D gains in PID) must first be defined. The trial and error, and the widely used Ziegler–Nichols tuning rules are usually employed to set initial gain values according to the open-loop step response of the plant. The ANN collects information about the system response and recommends adjustments to be made to the controller gains. This is an iterative procedure until the fastest possible critical damping for the controlled system is achieved. The main components of the ANN tuner include a response recognition unit to monitor the controlled response and extract knowledge about the performance of the current controller gain setting, and an embedded unit to suggest suitable changes to be made to the controller gains.

Combinations of ANN and robust control methodologies are presented for AGC synthesis in interconnected power systems.[80] These ideas use the robust performance indices provided by robust control techniques for desirable training of neural networks under various operating conditions. These approaches combine the advantage of neural networks and robust control techniques to achieve the desired level of robust performance under large parametric uncertainness and lead to a flexible controller with a relatively simple structure.

Application of ANN to AGC design is comprehensively presented in Chapter 5. The application is supplemented by some nonlinear simulation on single- and multiarea power system examples.

3.3 Genetic-Algorithm-Based AGC

A genetic algorithm (GA) is a searching algorithm that uses the mechanism of natural selection and natural genetics, operates without knowledge of the task domain, and utilizes only the fitness of evaluated individuals. The GA

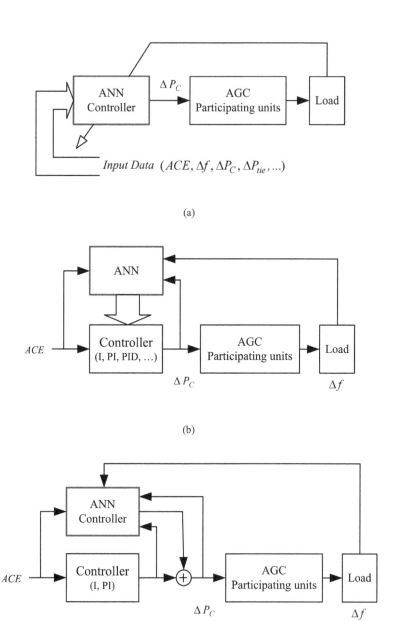

(a)

(b)

(c)

FIGURE 3.5
Common configurations for ANN-based AGC schemes in (a)–(c).

as a general purpose optimization method has been widely used to solve many complex engineering optimization problems over the years. In fact, GA as a random search approach that imitates the natural process of evolution is appropriate for finding a global optimal solution inside a multidimensional searching space. From the random initial population, GA starts a loop of evolution processes, consisting of selection, crossover, and mutation, in order to improve the average fitness function of the whole population.

Several reports are also available on the use of GAs in AGC systems.[41,81–90] GAs have been used to adjust parameters for different AGC schemes, e.g., integral, PI, PID, sliding mode control (SMC),[82,84,87,89,90] or variable structure control (VSC).[85,86] The overall control framework is shown in Figure 3.6.

An optimal adjustment of the classical AGC parameters for a two-area nonreheat thermal system using genetic techniques is investigated in Abdel-Magid and Dawoud.[81] A reinforced GA has been proposed as an appropriate optimization method to tune the membership functions, and rule sets for fuzzy gain scheduling of supplementary frequency controllers of multiarea power systems to improve the dynamic performance. The proposed control scheme incorporates dead-band and generation rate constraints also.

The GA in cooperation with fuzzy logic is used for optimal tuning of the integral controller's gain according to the performance indices' integral square error (ISE) and the integral of time multiplied by the absolute value of error (ITAE) in Chang et al.[84] A reinforced GA is employed to tune the membership function, and rule sets of fuzzy gain scheduling controllers to improve the dynamic performance of multiarea power systems in the presence of system nonlinearities, such as generation rate constraints (GRCs) and governor dead-band. Later, contrary to the trial-and-error selection of the variable-structure-based AGC feedback gains, a GA-based method was used for finding optimal feedback gains.[85] Parameters of a sliding mode

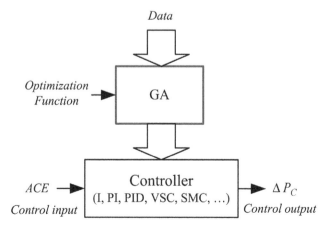

FIGURE 3.6
GA-based AGC scheme.

control-based AGC system are tuned in Vrdoljak et al.[90] using GA in a way to minimize the integral square of the area control error and control signal. Combinations of GA and robust control methods for AGC synthesis in a multiarea power system are shown in Rerkpreedapong et al.[41] The main idea is in running a GA optimization to track a robust performance index provided by robust control theory.

Chapter 11 addresses several GA-based AGC design methodologies and application examples. Chapter 11 shows the way to successfully use GAs for the tuning of control parameters, the solution of multiobjective optimization problems, satisfying robust performance indices, and the improvement of learning algorithms during the AGC synthesis process.

3.4 Multiagent-Based AGC

A multiagent system (MAS) is a system comprising two or more intelligent agents to follow a specific task. Nowadays, the MAS technology is being widely used in planning, monitoring, control, and automation systems. The MAS philosophy and its potential value to the power systems are discussed in McArthur et al.[91,92] Several reports have been published on the application of MAS technology with different characteristics and intelligent cores for the AGC systems.[25,93–98]

For a real-time AGC system, structural flexibility and having a degree of intelligence are highly important. In such systems, agents require real-time responses and must eliminate the possibility of massive communication among agents. In the synthesis of real-time MAS, the designer must at least denote the required number and type of agents in the system, the internal structure of each agent, and the communication protocol among the available agents.[25] Each agent is implemented on a software platform that supports the general components of the agent. The software platform must provide a communication environment among the agents and support a standard interaction protocol.

Some MAS-based frequency control scenarios using reinforcement learning, Bayesian networks, and GA intelligent approaches are addressed.[93,97,98] A multiagent reinforcement learning-based control approach with the capability of frequency bias estimation is proposed in Daneshfar and Bevrani.[94] It consists of two agents in each control area that communicate with each other to control the whole system. The first agent (estimator agent) provides the ACE following the area's frequency bias parameter estimation, and the second agent (controller agent) provides a control action signal according to the ACE signal received from the estimator agent, using reinforcement learning.

Bevrani[25] introduces an agent-based scenario to follow the main AGC objectives in a deregulated environment. An agent control system is used to cover a minimum number of required processing activities to the AGC objectives in a control area. The operating center for each area includes two agents: data acquisition/monitoring agent and decision/control agent. In the first agent, special care was taken regarding the provision of appropriate signals following a filtering and signal conditioning process on the measured signals and received data from the input channels. The sorted information and washout signals will be passed to the second agent. The decision/control agent uses the received data from the data acquisition/monitoring agent to provide the generation participation factors and appropriate control action signal, through an H∞-based robust PI controller. The decision/control agent evaluates the bids and performs the ACE signal using the measured frequency and tie-line power signals. The agent software estimates the total power imbalance and determines AGC participation factors, considering the ramp rate limits. The proposed control structure, which is summarized in Figure 3.7, is examined using real-time nonlinear simulation.

A multiagent-based frequency control scheme for isolated power systems with dispersed power sources such as photovoltaic units, wind generation units, diesel generation units, and an energy capacitor system (ECS) for the energy storage is presented.[95,96] The addressed scheme has been proposed through the coordination of controllable power sources such as the diesel units and the ECS with small capacity. All the required information for the proposed frequency control is transferred between the diesel units and the ECS through computer networks. A basic configuration of the proposed frequency regulation scheme is shown in Figure 3.8. In this figure, W_{ECS} and P_{ECS} are the current stored energy and produced power by the ECS unit, respectively. Experimental studies have been performed on the laboratory system to investigate the efficiency of the proposed multiagent-based control scheme.

In Chapter 7, the main frameworks for agent-based control systems are generally discussed. New multiagent-based AGC schemes are introduced, and the potential of reinforcement learning in the agent-based AGC systems is explained. Furthermore, an intelligent multiagent-based AGC design methodology using Bayesian networks is described in Chapter 8. The proposed approaches are supplemented by several simulations, including a real-time laboratory examination.

3.5 Combined and Other Intelligent Techniques in AGC

In the light of recent advances in artificial intelligent control and evolutionary computations, various combined intelligent control methodologies have been proposed to solve the power system AGC problem.[52,63,71,84,99–106] A study

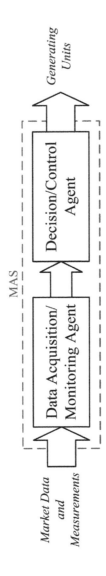

FIGURE 3.7
An agent-based AGC structure.

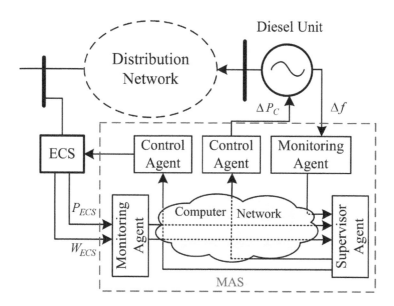

FIGURE 3.8
Agent-based frequency regulation scheme presented in Hiyama et al.[95,96]

on the AGC synthesis for an autonomous power system using the combined advent of ANN, fuzzy logic, and GA techniques is presented in Karnavas and Papadopoulos.[63] In Ghoshal,[99] a hybrid GA-simulated annealing (SA)-based fuzzy AGC scheme of a multiarea thermal generating system is addressed. A specific cost function named figure of demerit has been used as the fitness function for evaluating the fitness of GA/hybrid GA-SA optimization. This function directly depends on transient performance characteristics such as overshoots, undershoots, settling times, and time derivative of the frequency. The hybrid GA-SA technique yields more optimal gain values than the GA method.

For optimal tuning of PID gains in designing a Sugeno fuzzy-logic-based AGC scheme, a particle swarm optimization (PSO) technique was reported.[100] PSO, as one of the modern heuristic algorithms, is a population-based evolutionary algorithm that is motivated by simulation of social behavior instead of survival of the fittest. The proposed PSO algorithm establishes the true optimality of transient performance, similar to those obtained by the GA-SA-based optimization technique, but it is faster than the GA-SA algorithm. It has been shown in Ahamed et al.[104] that the AGC problem can be viewed as a stochastic multistage decision-making problem or a Markov chain control problem, and algorithms have been presented for designing AGC based on a reinforcement learning approach.

The reinforcement learning (RL) approach is used to AGC design a stochastic multistage decision problem in Ahamed et al.[104] Two specific RL-based AGC algorithms are presented. The first algorithm uses the traditional control objective of limiting ACE excursions, while in the second algorithm the controller can restore the load-generation balance by monitoring deviation in tie-line flows and system frequency without monitoring ACE signal. In Karnavas and Papadopoulos,[105] an intelligent load-frequency controller was developed using a combination of fuzzy logic, genetic algorithms, and neural networks to regulate the power output and system frequency by controlling the speed of the generator with the help of fuel rack position control.

Some heuristic stochastic search techniques such as GA, PSO, and bacteria foraging optimization have been proposed for optimization of PID gains used in Sugeno fuzzy-logic-based AGC of multiarea thermal generating plants.[103] An attempt is made to examine the application of bacterial foraging to optimize some AGC parameters in interconnected three unequal area thermal systems.[106] The parameters considered are integral controller gains for the secondary control, governor speed regulation parameters for the primary control, and frequency bias. The system performance is compared with the GA-based approach and classical methods.

3.6 AGC in a Deregulated Environment

In a traditional power system, generation, transmission, and distribution are owned by a single entity called a vertically integrated utility, which supplies power to the customers at regulated rates. Usually, the definition of a control area is determined by the geographical boundaries of the entity. Toward the end of the twentieth century many countries sought to reduce direct government involvement and, to increase economic efficiency, started to change the power system management structure, often described as *deregulation*.

There are several control scenarios and AGC schemes depending on the power system structure. Different organizations are introduced for the provision of AGC as an ancillary service in countries with restructured power systems. The AGC service and related transactions can be supervised by an independent system operator (ISO), independent contract administrator (ICA), transmission system operator (TSO), or another responsible organization. The type of AGC scheme in a restructured power system is differentiated by how free the market is, who controls generator units, and who has the obligation to execute the AGC.[107]

Several modeling and control strategies have been reported to adapt well-tested classical load-frequency control (LFC) schemes to the changing environment of power system operation under deregulation.[25,108–113] A generalized modeling structure based on the introduced idea in several references[25,112,114] is presented in Chapter 4. The effects of deregulation of the power industry on AGC, several AGC schemes, and control scenarios for power systems after deregulation have been addressed.[25,107,113–121]

An agent-based AGC scenario in a deregulated environment is introduced in Bevrani.[25] In order to cover a minimum number of required processing activities for the AGC objectives in a control area, a two-agent control system is used. Based on the proposed control strategy, a decision and control agent uses the received data from a data acquisition and monitoring agent to provide the generation participation factors and appropriate control action signal, through a simple robust controller. The system frequency has been analytically described, and to demonstrate the efficiency of the proposed control method, real-time nonlinear laboratory tests have been performed.

A decentralized robust AGC design in a deregulated power system under a bilateral-based policy scheme is addressed in Bevrani et al.[120] In each control area, the effect of bilateral contracts is taken into account as a set of new input signals in a modified traditional dynamical model. The AGC problem is formulated as a multiobjective control problem via a mixed H2/H∞ control technique. In order to design a robust PI controller, the control problem is reduced to a static output feedback control synthesis, and then it is solved using a developed iterative linear matrix inequalities algorithm to get a robust performance index close to a specified optimal one. The results of the proposed multiobjective PI controllers are compared with H2/H∞ dynamic controllers.

Chapter 4 reviews the main structures, characteristics, and existing challenges for AGC synthesis in a deregulated environment. The AGC response and an updated AGC mode concerning the bilateral contracts are also introduced in Chapter 4.

3.7 AGC and Renewable Energy Options

The power system is currently undergoing fundamental changes in its structure. These changes are associated not just with the deregulation issue and the use of competitive policies, but also with the use of new types of power production, new technologies, and rapidly increasing amounts of renewable energy sources (RESs).

The increasing need for electrical energy, as well as limited fossil fuel reserves, and the increasing concerns with environmental issues call for fast development in the area of RESs. Renewable energy is derived from natural sources such as the sun, wind, hydropower, biomass, geothermal, and oceans. These changes imply a requirement for new AGC schemes in modern power systems. Recent studies have found that the renewable integration impacts on system frequency and power fluctuation are nonzero and become more significant at higher sizes of penetrations. The variability and uncertainty are two major attributes of variable RESs that notably impact the bulk power system planning and operations. The main difficulties are caused by the variability and limited predictability of power from renewable sources such as wind, PV, and waves.

Integration of RESs into power system grids has impacts on optimum power flow, power quality, voltage and frequency control, system economics, and load dispatch. The effects of variability are different than the effects of uncertainty, and the mitigation measures that can be used to address each of these are different. Regarding the nature of RES power variation, the impact on the frequency regulation issue has attracted increasing research interest during the last decade. Some studies represent a range of estimates based on different system characteristics, penetration levels, and study methods.[25] The RESs have different operational characteristics relative to the traditional forms of generating electric energy, and they affect the dynamic behavior of the power system in a way that might be different from conventional generators. This is due to the fact that the primary energy source in most RES types (such as wind, sunlight, and moving water) cannot presently be controlled or stored. Unlike coal or natural gas, which can be extracted from the earth, delivered to power plants thousands of miles away, and stockpiled for use when needed, variable fuels must be used when and where they are available. This fact alone results in RESs being viewed as nondispatchable, implying it is not possible to a priori specify what the power output of a RES unit should be.

3.7.1 Present Status and Future Prediction

RESs' revolution has already commenced in many countries, as evidenced by the growth of RESs in response to the climate change challenge and the need to enhance fuel diversity. Renewable energy currently provides more than 14% of the world's energy supply.[122]

Currently, wind is the most widely utilized renewable energy technology in power systems, and its global production is predicted to grow to more than 300 GW in 2015. It has been predicted that wind power global penetration will reach 8% by 2020. The European Union has set as a target 20% of electricity supplied by renewable generation by 2020.[123] According to the European Wind Energy Association (EWEA), European wind power capacity is expected to be more than 180 GW in 2020.[124] The U.S. Department of Energy has announced a goal of obtaining 6% of U.S. electricity only from wind by 2020—a goal that is consistent with the current growth rate of wind energy nationwide.[125]

Numerous works on solar (PV) energy, batteries, and energy capacitor units are being performed in Japan. Japan has set the target PV installed capacity of 28 GW by the year 2010.[126] The growing wind power market in Asian countries is also impressive. Based on current growth rates, the Chinese Renewable Energy Industry Association (CREIA) forecasts a capacity of around 50,000 MW by 2015.[127] India also continues to see a steady growth in wind power installations. In Korea, RES is gradually growing per year, and the government plans to replace 5% of the conventional energy source by the year 2011.[123]

After some slow years, the Pacific market has gained new impetus. In Australia, the government has pledged to introduce a 20% target for renewable energy by 2020. Although Europe, North America, Asia, and the Pacific region continue to have the largest additions to their RES capacity, the Middle East, North Africa, and Latin America have also increased their RES installations. New capacity was mostly added in Iran, Egypt, Morocco, Tunisia, and Brazil.[127]

3.7.2 New Technical Challenges

High renewable energy penetration in power systems may increase uncertainties during abnormal operation, introduce several technical implications, and open important questions as to whether the traditional power system control approaches (such as AGC systems) to operation in the new environment are still adequate. The main question that arises is: What happens to the AGC requirements if numerous RESs are added to the existing generation portfolio?

The impacts of large-scale penetration of variable RESs should be considered in terms of appropriate timeframes. In the seconds-to-minutes timeframe, overall power system reliability is almost entirely controlled by automatic equipment and control systems such as AGC systems, generator governor and excitation systems, power system stabilizers (PSSs), automatic voltage regulators (AVRs), protective relaying and special protection schemes, and fault ride-through capability of the generation resources. From the minutes-to-one-week timeframe, system operators and operational planners must be able to commit or dispatch needed facilities to rebalance, restore, and position the whole power system to maintain reliability through normal load variations as well as contingencies and disturbances.

The RES units must meet technical requirements with respect to voltage, frequency, and ability to rapidly isolate faulty parts from the rest of the network, and have a reasonable ability to withstand abnormal system operating conditions. They should be able to function effectively as part of the existing electricity industry, particularly during abnormal power system operating conditions when power system security may be at risk. High RES penetration, particularly in locations far away from major load centers and existing conventional generation units, increases the risk of tie-line overloading, and may require network augmentation, and possibly additional interconnections to avoid flow constraints. With the increasing RES penetration, the grid code for the connection of high RES capacity should be also updated.

Recent investigation studies indicate that relatively large-scale wind generation will have an impact on power system frequency regulation and AGC systems, as well as other control and operation issues. This impact may increase at penetration rates that are expected over the next several years. On the other hand, most existing variable RESs today do not have the control capability necessary to provide regulation. But perhaps even more significant is that the variability associated with those energy sources not only does not help regulate, but it contributes to a need for more regulation. The AGC system of the future will require increased flexibility and intelligence to ensure that it can continuously balance fluctuating power and regulate frequency deviation caused by renewable energy sources such as wind, solar, or wave power.

To maintain reliable and efficient operation of the power system, operators must use forecasts of demand and generator availability. Today the majority of supply-demand balancing in a power system is achieved by controlling the output of dispatchable generation resources to follow the changes in demand. Typically, a smaller portion of the generation capacity in a control area is capable of and designated to provide AGC service in order to deal with the more rapid and uncertain demand variations, often within the seconds-to-minutes timeframe. AGC is expected to play a major role in managing short-term uncertainty of variable renewable power, and to mitigate some of the short-term impacts associated with variable generation forecast error. Hence, it may be necessary for planners and operators to review and potentially modify the AGC performance criteria, capabilities, and technologies to ensure that these systems perform properly.[128]

In response to the above-mentioned challenges, intelligent control certainly plays a significant role. It will not be possible to integrate large amounts of RESs into the conventional power systems without intelligent control and regulation systems. For this purpose, intelligent meters, devices, and communication standards should first be prepared to enable flexible matching of generation and load. Furthermore, an appropriate framework must be developed to ensure that future flexible supply/demand and ancillary services have equal access and are free to the market.

3.7.3 Recent Achievements

Here, a brief critical literature review and an up-to-date bibliography for the proposed studies on the frequency, tie-line power flow, and AGC issue in the presence of RESs, and associated issues, are presented. A comprehensive survey on past achievements and open research issues is presented in Bevrani et al.[129]

A considerable part of attempts has focused on wind power generation units. Integrating energy storage systems (ESSs) or energy capacitor systems (ECSs) into the wind energy system to diminish the wind power impact on power system frequency has been addressed in several reported works.[25,130] Different ESSs by means of an electric double-layer capacitor (EDLC) and SMES and energy saving are proposed for wind power leveling. The impact of wind generation on the operation and development of the UK electricity systems is described in Strbac et al.[131] Impacts of wind power components and variations on power system frequency control are described in Banakar et al.[132] and Lalor et al.[133] Using the kinetic energy storage system (blade and machine inertia) to participate in primary frequency control is addressed in Morren et al.[134]

The technology to filter out the power fluctuations (in resulting frequency deviation) by wind turbine generators for the increasing amount of wind power penetration is growing. The new generation of variable speed, large wind turbine generators with high moments of inertia from their long turbine blades can filter power fluctuations in the wind farms. A method is presented in Morren et al.[134] to let variable speed wind turbines emulate inertia and support primary frequency control.

A method of quantifying wind penetration based on the amount of fluctuating power that can be filtered by wind turbine generation and thermal plants is addressed in Luo et al.[135] A small power system including three thermal units (equipped with an AGC system) and a wind farm is considered as a test example. Using the Bode diagram of system transfer function between frequency deviation and real power fluctuation signals, the permitted power fluctuation for 1% frequency deviation is approximated.

Using modal techniques, the dynamic influence of wind power on the primary and supplementary frequency controls is studied.[136,137] Some preliminary studies showed that the kinetic energy stored in the rotating mass of a wind turbine can be used to support primary frequency control for a short period of time.[134] The capability of providing a short-term active power support of a wind farm to improve the primary frequency control performance is discussed in Ullah et al.[138]

Some recent studies analyze the impacts of RESs on AGC operation and supplementary frequency control.[133,137,139–143] A study is conducted in Hirst[140] to help determine how wind generation might interact in the competitive wholesale market for regulation services and a real-time balancing market. This study recognized that wind integration does not require that each deviation in wind power output be matched by a corresponding and opposite deviation in other resources, and the frequency performance requirement must apply to the aggregated system, not to each individual generator. Several works are reported on considering the effect of wind power fluctuation on LFC structure.[133] An automatic generation control system for a wind farm with variable speed turbines is addressed in Amenedo et al.[139] The proposed integrated control system includes two control levels (supervisory system and machine control system). Several multiagent, ANN, and fuzzy-logic-based intelligent control schemes for AGC systems concerning the wind farms are given.[137,143]

The impacts of wind power on tie-line power flow in the form of low-frequency oscillations due to insufficient system damping are studied in Slootweg and Kling[144] and Chompoo-inwai et al.[145] The need for retuning of UFLS df/dt relays is emphasized in Bevrani and coworkers.[25,129] Using storage devices such as ESSs, ECSs, and redox flow (RF) batteries for supplementary control and maintenance of power quality in the presence of distributed power resources is suggested in many published works. It has been shown that the LFC capacity of RF battery systems is ten times that of fossil power systems, due to quick response characteristics.

Recently, several studies have been conducted on the required regulation reserve estimation in the presence of various RES units. Some efforts to evaluate the impact of small PV power generating stations on economic and performance factors for a large-scale power system are addressed in Asano et al.[126] It was found that wind power, combined with the varying load, does not impose major extra variations on the system until a substantial penetration is reached. Large geographical spreading of wind power will reduce variability, increase predictability, and decrease the occasions with near zero or peak output. It is investigated in Holttinen[142] that the power fluctuation from geographically dispersed wind farms will be uncorrelated with each other, hence smoothing the sum power and not imposing any significant requirement for additional frequency regulation reserve, and required extra balancing is small.

Chapter 6 addresses the AGC system concerning the integration of RES. It presents the impact of power fluctuation produced by variable RESs on the AGC performance, and gives an updated AGC frequency response model. The statements are supported by some nonlinear time-domain simulations on standard test systems.

3.8 AGC and Microgrids

A microgrid is a relatively novel concept in modern electric industry, consisting of small power systems owning the capability of performing isolated from the main network. A microgrid can tackle all distributed energy resources, including distributed generation (DG), RESs, distributed energy storage systems, and demand response, as a unique subsystem, and offers significant control capacities on its operation. Microgrids are usually based on loads fed through a low- or medium-voltage level, mostly in distribution radial systems. In order to ensure proper operation of the microgrid, it is important that its constituent parts and controls in both grid-connected and islanded modes operate satisfactorily.

A schematic diagram of a generic microgrid is shown in Figure 3.9. The microgrid is connected to the main grid at the point of interconnection (POI). The DG units of the microgrid can in general have any arbitrary configuration. Each DG unit is usually interfaced to the microgrid through a power electronic converter, at the point of coupling (POC). As mentioned, many of the DGs/RESs, such as PV and fuel cells, use a power electronic converter (inverter) for grid interfacing. Some wind applications as well as some synchronous machines and micro-turbines utilize power electronic devices for the grid interface, as the benefits of the electronic interface justify the additional cost and complexity.

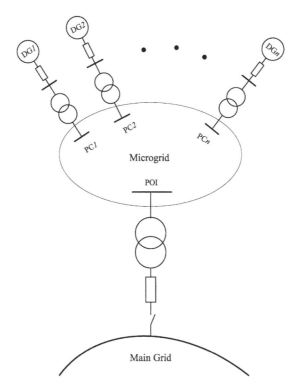

FIGURE 3.9
A simplified generic microgrid.

Power electronic inverters are capable of converting the energy from a variety of sources, such as variable frequency (wind), high frequency (turbines), and direct energy (PV and fuel cells). Inverter-based DGs/RESs are generally considered low power by utility standards, from 1 kW up to a few MW. Generators connected to renewable sources are not reliable, and so are not considered dispatchable by the utility and are not tightly integrated into the power supply system. The inverter interface decouples the generation source from the distribution network. Since inverters monitor the frequency at their output terminals for control purposes, it is easy to detect when the inverter frequency shifts outside a window centered on the nominal frequency set point.[129]

Increases in distributed generation, microgrids, and active control of consumption open the way to new control strategies with a greater control hierarchy/intelligence and decentralized property. In this direction, recently several new concepts and national projects, such as SmartGrids,[146] Intelligrid,[147] and Gridwise,[148] have been defined in Europe and the United States. Their aim has been to take advantage of the possibilities created by combining intelligent communication, information technology, and distributed energy resources in the power system.

Similar to conventional generating units, droop control is one of the important control methods for a microgrid with multiple DG units. Without needing a communication channel and specific coordination, the DG units can automatically adjust their set points using the frequency measurement to meet the overall need of the microgrid.[149]

Unlike large power systems, the drooping system is poorly regulated in microgrids to support spinning reserve as an ancillary service in power markets. Some recent works address the scheduling of the droop coefficients for frequency regulation in microgrids.[150,151] A methodology based on bifurcation theory is used in Chandorkar et al.[151] to evaluate the impact that droops and primary reserve scheduling have on the microgrid stability. It was demonstrated in Diaz et al.[150] that drooping characteristics can be successfully applied for controlling paralleled inverters in isolated alternative current (AC) systems, mimicking the performance of generator-turbine-governor units.

On the other hand, tie-line control can be designed to manage the feeder power flow at the POI to meet the needs of the system operator. Control is implemented by coordinating the assets of the microgrid, allowing the collection of these assets to appear as one aggregated dispatchable producing or consuming entity connected to the main grid. The overall objective is to optimize operating performance and cost in the normally grid-connected mode, while ensuring that the system is capable of meeting the performance requirements in stand-alone mode.[152] Enforcing power and ramp rate limits at the POI with appropriate responses to system frequency deviation can be achieved via microgrid active power control. It is noteworthy that unlike standard AGC implementation, tie-line control is not applicable when an isolated microgrid is considered.

Another major issue in the area of microgrid control is the move toward active control of domestic loads. However, the flexibility introduced through the ability to control part of household power consumption is difficult to exploit, and up to now there have been no significant achievements.

As mentioned above, microgrids and DGs have great potential in contributing to the frequency control of the overall system. The main challenge is to coordinate their actions so that they can provide the regulation services. For this purpose, the aggregation technique to create *virtual power plants* (VPPs) could be useful. In this method, the aggregation is not based on the topology of the network; instead, any unit can participate in a VPP, regardless of its location. It facilitates the provision of some services (by VPPs), such as frequency regulation and power balance, which are global characteristics.

The possibility of having numerous controllable DG units and microgrids in distribution networks requires the use of intelligent and hierarchical control schemes that enable efficient control and management of this kind of system. During the last few years, several reports presenting various control methods on frequency regulation, tie-line control, and LFC/AGC issues have been published.[152-158] A PI control, following estimation of power demand (amount of load disturbance), is used in Fujimoto et al.[153] The method is applied on a real microgrid, including PV, micro-hydropower generator, diesel generator, storage system, and wind turbine. The tie-line controls are important microgrid control features that regulate the active and reactive power flow between the microgrid and the main grid at the point of interconnection, and are addressed in Adamiak et al.[152] These controls essentially allow the microgrid to behave as an aggregated power entity that can be made dispatchable by the utility. Molina and Mercado[154] have used a combination of SMES and a distribution static synchronous compensator to control the tie-line power flow of microgrids. An electrolyzer system with a fuzzy PI control is used in Li et al.[159] to solve power quality issues resulting from microgrid frequency fluctuations.

Fuzzy logic PID controllers for frequency regulation in isolated microgrids are given in Chaiyatham et al.[155] and Hiyama et al.[158] A bee colony optimization is used in Chaiyatham et al.[155] to simultaneously optimize scale factors, membership function, and control rules of the fuzzy controller. A general and a hierarchical frequency control scheme for isolated microgrid operation are addressed in Madureia et al.[156] and Gil and Pecas Lopes,[157] respectively.

3.9 Scope for Future Work

Restructuring and introducing new uncertainty and variability by a significant number of DGs, RESs, and microgrids into power systems add new economical and technical challenges associated with AGC systems synthesis and analysis. As the electric industry seeks to reliably integrate large amounts of variable generation into the bulk power system, considerable effort will be needed to accommodate and effectively manage these unique operating and planning characteristics. A key aspect is how to handle changes in topology caused by switching in the network, and how to make the AGC system robust and able to take advantage of the potential flexibility of distributed energy resources.

3.9.1 Improvement of Modeling and Analysis Tools

A complete understanding of reliability considerations via effective modeling/aggregation techniques is vital to identify a variety of ways that bulk power systems can accommodate the large-scale integration of DGs/RESs. A more complete dynamic frequency response model is needed in order to analyze and synthesize AGC in interconnected power systems with a high degree of DG/RES penetration. Proper dynamic modeling and aggregation of the distributed generating units, for AGC studies, is a key issue to understand the dynamic impact of distributed resources and simulate the AGC functions in new environments.

3.9.2 Develop Effective Intelligent Control Schemes for Contribution of DGs/RESs in the AGC Issue

Additional flexibility may be required from various dispatchable generators, storage, and demand resources, so the system operator can continue to balance supply and demand on the modern bulk power system. The contribution of DGs/RESs (microgrids) in the AGC task refers to the ability of these systems to regulate their power output, either by disconnecting a part of generation or by an appropriate control action. More effective practical algorithms and control methodologies are needed to address these issues properly. Since the power coming from some DGs/RESs, specifically wind turbine, is stochastic, it is difficult to straightly use their kinetic energy storage in AGC. Further studies are needed to coordinate the timing and size of the kinetic energy discharge with the characteristics of conventional generating plants.

3.9.3 Coordination between Regulation Powers of DGs/RESs and Conventional Generators

In the case of supporting the AGC system, an important feature of some DG/RES units is the possibility of their fast active power injection. Following a power imbalance, the active power generated by DG/RES units quickly changes to recover the system frequency. Because this increased/decreased power can last just for a few seconds, conventional generators should eventually take charge of the huge changed demand by shifting their generation to compensate power imbalance.[160] But the fast power injection by DG/RES units may slow down, to a certain extent, the response of conventional generators. To avoid this undesirable effect, coordination between DG/RES and conventional participating units in the AGC system is needed.

3.9.4 Improvement of Computing Techniques and Measurement Technologies

The AGC system of tomorrow must be able to handle complex interactions between control areas, grid interconnections, distributed generating equipment and RESs, fluctuations in generating capacity, and some types of controllable demand, while maintaining security of supply. These efforts are directed at developing computing techniques, intelligent control, and monitoring/measurement technologies to achieve optimal performance. Advanced computational methods for predicting prices, congestion, consumption, and weather, and improved measuring technologies are opening up new ways of controlling the power system via supervisory control and data acquisition (SCADA)/AGC centers.

The design of a definitive frequency threshold detector and trigger to reduce/increase the contribution of DG/RES generators requires extensive research to incorporate signal processing, adaptive strategies, pattern recognition, and intelligent features to achieve the same primary reserve capability of conventional plants. An advanced computing algorithm and fast hardware measurement devices are also needed to realize optimal/adaptive AGC schemes for modern power systems.

3.9.5 Use of Advanced Communication and Information Technology

The AGC structure for future power systems must be able to control/regulate frequency and power, and monitor itself as an intelligent system to a greater extent than is the case today. Key components in the intelligent AGC system of the future will thus be systems for metering, controlling, regulating, and monitoring indices, allowing the resources of the power system to be used effectively in terms of both economics and operability. To achieve this vision, the future AGC systems must include advanced communications and information technology (IT).

Continuous development of metering, communications, market frameworks, and regulatory frameworks for generation and consumption is a precondition for a power system with intelligent electricity meters and intelligent communications.[161]

3.9.6 Update/Define New Grid Codes

Further study is needed to define new grid codes for the contribution of microgrid, large DG/RES units (connected to the transmission system) to the AGC issue, and for investigation of their behavior in the case of abnormal operating conditions in the electric network. The active power ramp rate must comply with the respective rates applicable to conventional power units.[162] The new grid codes should clearly impose the requirements on the regulation capabilities of the active power produced by microgrids and distributed sources.

3.9.7 Revising of Existing Standards

Standards related to the overall reliable performance of the power system as instituted by technical committees, reliability entities, regulatory bodies, and organizations (such as NERC, UCTE, ISOs, and RTOs, etc.) ensure the integrity of the whole power system is maintained for credible contingencies and operating conditions. There exist some principles to be taken into account in future standards development on the AGC system in the presence of DGs/RESs and microgrids. Standards should be comprehensive, transparent, and explicit to avoid misinterpretation. Interconnection procedures and standards should be enhanced to address frequency regulation, real power control, and inertial response, and must be applied in a consistent manner to all generation technologies.

The overall behavior expected from a power system with high levels of variable generation will be different from what is experienced today; therefore, both the equipment design and performance requirements must be addressed. In this respect, reliability-focused equipment standards must also be further developed to facilitate the reliable integration of additional DGs/RESs into the bulk power system. From a bulk power system reliability perspective, a set of interconnection procedures and standards is required that applies equally to all generation resources interconnecting to the power grid. Significant work is required to standardize basic requirements in these interconnection procedures and standards, such as the ability of the generator owner and operator to provide an inertial response (effective inertia as seen from the grid), control of the MW ramp rates or curtailing of MW output, and frequency control (governor action, AGC, etc.).[128]

The requirements imposed should reflect an optimum balance between cost and technical performance. Proper consideration should also be given to coordinate the large-scale interconnected systems, via their responsible control centers and organizations, such as the collaboration among neighboring transmission system operators (TSOs) and the AGC owner. Finally, frequency performance standards compliance verification remains a major open issue for DG/RES units.

3.9.8 Updating Deregulation Policies

To allow for increased penetration of DGs/RESs and microgrids, a change in regulation reserve policy may be required. In this direction, in addition to deregulation policies, the amount and location of distributed sources, generation technology, and the size and characteristics of the electricity system must be considered as important technical aspects. Moreover, the updating of existing emergency frequency control schemes for $N - 1$ contingency, concerning economic assessment/analysis, the frequency of regulation prices, and other economical, social, and political issues, and the quantification of a reserve margin due to increasing DG/RES penetration are some important research needs.

3.10 Summary

The AGC design problem in power systems can be easily transferred into a performance optimization problem, which is suitable for application of artificial intelligent techniques. This has led to emerging trends of application of soft computing or computational intelligence and evolutionary computing in the power system AGC synthesis issue.

In this chapter, the most important issues on the intelligent AGC with the past achievements in this literature are briefly reviewed. The most important intelligent AGC design frameworks based on fuzzy logic, neural network, neuro-fuzzy, genetic algorithm, multiagent system, and other intelligent techniques and evolutionary optimization approaches are described. An overview of the key issues in the power system intelligent AGC—deregulation and integration of RESs/DGs and microgrids—is presented. The need for further research on the intelligent AGC and related areas, including the necessity of revising existing performance standards, is emphasized.

References

1. C. Concordia, L. K. Kirchmayer. 1953. Tie line power and frequency control of electric power systems. *Am. Inst. Elect. Eng. Trans.* 72(Part II):562–72.
2. N. Cohn. 1957. Some aspects of tie-line bias control on interconnected power systems. *Am. Inst. Elect. Eng. Trans.* 75:1415–36.
3. L. K. Kirchmayer. 1959. *Economic control of interconnected systems.* New York: Wiley.
4. J. E. Van Ness. 1963. Root loci of load frequency control systems. *IEEE Trans. Power App. Syst.* PAS-82(5):712–26.
5. IEEE. 1970. Standard definitions of terms for automatic generation control on electric power systems. IEEE committee report. *IEEE Trans. Power App. Syst.* PAS-89:1356–64.
6. O. I. Elgerd, C. Fosha. 1970. Optimum megawatt-frequency control of multiarea electric energy systems. *IEEE Trans. Power App. Syst.* PAS-89(4):556–63.
7. C. Fosha, O. I. Elgerd. 1970. The megawatt-frequency control problem: A new approach via optimal control. *IEEE Trans. Power App. Syst.* PAS-89(4):563–77.
8. IEEE. 1979. Current operating problems associated with automatic generation control. IEEE PES committee report. *IEEE Trans. Power App. Syst.* PAS-98:88–96.
9. N. Jaleeli, D. N. Ewart, L. H. Fink. 1992. Understanding automatic generation control. *IEEE Trans. Power Syst.* 7(3):1106–12.
10. C. Concordia, L. K. Kirchmayer, E. A. Szymanski. 1957. Effect of speed governor dead-band on tie-line power and frequency control performance. *Am. Inst. Elect. Eng. Trans.* 76:429–35.
11. F. F. Wu, V. S. Dea. 1978. Describing-function analysis of automatic generation control system with governor deadband. *Elect. Power Syst. Res.* 1(2):113–16.
12. S. C. Tripathy, T. S. Bhatti, C. S. Jha, O. P. Malik, G. S. Hope. 1984. Sampled data automatic generation control analysis with reheat steam turbines and governor dead band effects. *IEEE Trans. Power App. Syst.* AS-103(5):1045–51.
13. T. Hiyama. 1982. Optimisation of discrete-type load-frequency regulators considering generation-rate constraints. *IEE Proc. C* 129(6):285–89.
14. L. D. Douglas, T. A. Green, R. A. Kramer. 1994. New approaches to the AGC nonconforming load problem. *IEEE Trans. Power Syst.* 9(2):619–28.
15. L. D. Douglas, T. A. Green, R. A. Kramer. 1994. New approach to the AGC nonconforming load program. *IEEE Trans. Power Syst.* 9(2):619–28.

16. R. P. Schulte. 1996. An automatic generation control modification for present demands on interconnected power systems. *IEEE Trans. Power Syst.* 11(3): 1286–94.

17. N. Jaleeli, L. S. Vanslyck. 1999. NERC's new control performance standards. *IEEE Trans. Power Syst.* 14(3):1092–99.

18. M. Yao, R. R. Shoults, R. Kelm. 2000. AGC logic based on NERC's new control performance standard and disturbance control standard. *IEEE Trans. Power Syst.* 15(2):852–57.

19. Y. Hain, R. Kulessky, G. Nudelman. 2000. Identification-based power unit model for load-frequency control purposes. *IEEE Trans. Power Syst.* 15(4):1313–21.

20. N. Maruejouls, T. Margotin, M. Trotignon, P. L. Dupuis, J. M. Tesseron. 2000. Measurement of the load frequency control system service: Comparison between American and European indicators. *IEEE Trans. Power Syst.* 15(4):1382–87.

21. G. Gross, J. W. Lee. 2001. Analysis of load frequency control performance assessment criteria. *IEEE Trans. Power Syst.* 16(3):520–32.

22. N. Hoonchareon, C. M. Ong, R. A. Kramer. 2002. Feasibility of decomposing ACE_1 to identify the impact of selected loads on CPS1 and CPS2. *IEEE Trans. Power Syst.* 17(3):752–56.

23. L. R. C. Chien, C. M. Ong, R. A. Kramer. 2003. Field tests and refinements of an ACE model. *IEEE Trans. Power Syst.* 18(2):898–903.

24. L. R. C. Chien, N. Hoonchareon, C. M. Ong, R. A. Kramer. 2003. Estimation of β for adaptive frequency bias setting in load frequency control. *IEEE Trans. Power Syst.* 18(2):904–11.

25. H. Bevrani. 2009. *Robust power system frequency control.* New York: Springer.

26. Ibraheem, P. Kumar, P. Kothari. 2005. Recent philosophies of automatic generation control strategies in power systems. *IEEE Trans. Power Syst.* 20(1):346–57.

27. H. Glavitsch, J. Stoffel. 1980. Automatic generation control. *Elect. Power Energy Syst.* 2(1):21–28.

28. E. V. Bohn, S. M. Miniesy. 1972. Optimum load frequency sample data control with randomly varying system disturbances. *IEEE Trans. Power App. Syst.* PAS-91(5):1916–23.

29. K. Yamashita, T. Taniguchi. 1986. Optimal observer design for load frequency control. *Int. J. Elect. Power Energy Syst.* 8(2):93–100.

30. M. S. Calovic. 1972. Linear regulator design for a load and frequency control. *IEEE Trans. Power App. Syst.* 91:2271–85.

31. T. Hiyama. 1982. Design of decentralised load-frequency regulators for interconnected power systems. *IEE Proc. C* 129(1):17–23.

32. J. Kanniah, S. C. Tripathy, O. P. Malik, G. S. Hope. 1984. Microprocessor-based adaptive load-frequency control. *IEE Proc. Gener. Transm. Distrib.* 131(4):121–28.

33. A. Feliachi. 1987. Optimal decentralized load frequency control. *IEEE Trans. Power Syst.* PWRS-2(2):379–84.

34. O. P. Malik, A. Kumar, G. S. Hope. 1988. A load frequency control algorithm based on a generalized approach. *IEEE Trans. Power Syst.* 3(2):375–82.

35. V. R. Moorthi, P. P. Aggarwal. 1982. Suboptimal and near optimal control of a load-frequency control system. *IEE Proc. C* 129(6):1635–60.

36. A. Y. Sivaramakrishnan, M. V. Hartiharan, M. C. Srisailam. 1984. Design of variable structure load frequency controller using pole assignment technique. *Int. J. Control* 40:487–98.

37. A. Z. Al-Hamouz, Y. L. Al-Magid. 1993. Variable structure load frequency controllers for multiarea power systems. *Elect. Power Energy Syst.* 15:293–300.

38. W. C. Chan, Y. Y. Hsu. 1981. Automatic generation control of interconnected power systems using variable-structure controller. *Proc. Inst. Elect. Eng. C* 128(5):269–79.

39. A. Kumar, O. P. Malik, G. S. Hope. 1985. Variable-structure-system control applied to AGC of an interconnected power system. *Proc. Inst. Elect. Eng. C* 132(1):23–29.

40. H. Bevrani, T. Hiyama. 2009. On robust load-frequency regulation with time delays: Design and real-time implementation. *IEEE Trans. Energy Conversion* 24(1):292–300.

41. D. Rerkpreedapong, A. Hasanovic, A. Feliachi. 2003. Robust load frequency control using genetic algorithms and linear matrix inequalities. *IEEE Trans. Power Syst.* 18(2):855–61.

42. H. Bevrani, Y. Mitani, K. Tsuji. 2004. Robust decentralized load-frequency control using an iterative linear matrix inequalities algorithm. *IEE Proc. Gener. Transm. Distrib.* 151(3):347–54.

43. H. Bevrani, T. Hiyama. 2007. Robust load-frequency regulation: A real-time laboratory experiment. *Optimal Control Appl. Methods* 28(6):419–33.

44. I. Vajk, M. Vajta, L. Keviczky, R. Haber, J. Hetthessy, K. Kovacs. 1985. Adaptive load frequency control of the Hungarian power system. *Automatica* 21:129–37.

45. C. T. Pan, C. M. Liaw. 1989. An adaptive controller for power system load frequency control. *IEEE Trans. Power Syst.* PWRS-4:122–28.

46. A. Rubaai, V. Udo. 1994. Self-tuning load frequency control: Multilevel adaptive approach. *IEE Proc. Gener. Transm. Distrib.* 141(4):285–90.

47. L. M. Smith, L. H. Fink, R. P. Schulz. 1975. Use of computer model of interconnected power system to assess generation control strategies. *IEEE Trans. Power App. Syst.* 94(5).

48. M. L. Kothari, J. Nanda, D. P. Kothari, D. Das. 1989. Discrete mode automatic generation control of a two area reheat thermal system with new area control error. *IEEE Trans. Power App. Syst.* 4(2):730–38.

49. I. Ngamroo, Y. Mitani, K. Tsuji. 1999. Application of SMES coordinated with solid-state phase shifter to load frequency control. *IEEE Trans. Appl. Superconductivity* 9(2):322–25.

50. Y. Yoshida, T. Machida. 1969. Study of the effect of the DC link on frequency control in interconnected AC systems. *IEEE Trans. Power App. Syst.* PAS-88(7):1036–42.

51. M. Sanpei, A. Kakehi, H. Takeda. 1994. Application of multi-variable control for automatic frequency controller of HVDC transmission system. *IEEE Trans. Power Del.* 9(2):1063–68.

52. C. F. Juang, C. F. Lu. 2006. Load-frequency control by hybrid evolutionary fuzzy PI controller. *IEE Proc. Gener. Transm. Distrib.* 153(2):196–204.

53. C. S. Chang, W. Fu. 1997. Area load frequency control using fuzzy gain scheduling of PI controllers. *Elect. Power Syst. Res.* 42:145–52.

54. G. A. Chown, R. C. Hartman. 1998. Design and experience with a fuzzy logic controller for automatic generation control (AGC). *IEEE Trans. Power Syst.* 13(3):965–70.

55. J. Talaq, F. Al-Basri. 1999. Adaptive fuzzy gain scheduling for load frequency control. *IEEE Trans. Power Syst.* 14(1):145–50.

56. A. Feliachi, D. Rerkpreedapong. 2005. NERC compliant load frequency control design using fuzzy rules. *Elect. Power Syst. Res.* 73:101–6.

57. A. Demiroren, E. Yesil. 2004. Automatic generation control with fuzzy logic controllers in the power system including SMES units. *Elect. Power Energy Syst.* 26:291–305.

58. M. K. El-Sherbiny, G. El-Saady, A. M. Yousef. 2002. Efficient fuzzy logic load-frequency controller. *Energy Conversion Management* 43:1853–63.

59. E. Yesil, M. Guzelkaya, I. Eksin. 2004. Self tuning fuzzy PID type load frequency controller. *Energy Conversion Management* 45:377–90.

60. C. S. Indulkar, B. Raj. 1995. Application of fuzzy controller to automatic generation control. *Elect. Machines Power Syst.* 23(2):209–20.

61. A. E. Gegov, P. M. Frank. 1995. Decomposition of multivariable systems for distributed fuzzy control [power system load frequency control]. *Fuzzy Sets Syst.* 73(3):329–40.

62. E. Cam, I. Kocaarslan. 2005. A fuzzy gain scheduling PI controller application for an interconnected electrical power system. *Elect. Power Syst. Res.* 73:267–74.

63. Y. L. Karnavas, D. P. Papadopoulos. 2002. AGC for autonomous power system using combined intelligent techniques. *Elect. Power Syst. Res.* 62:225–39.

64. C. S. Rao, S. S. Nagaraju, P. S. Raju. 2009. Automatic generation control of TCPS based hydrothermal system under open market scenario: A fuzzy logic approach. *Elect. Power Energy Syst.* 31:315–22.

65. A. P. Fathima, M. A. Khan. 2008. Design of a new market structure and robust controller for the frequency regulation service in the deregulated power system. *Elect. Power Components Syst.* 36:864–83.

66. P. Subbaraj, K. Manickavasagam. 2008. Automatic generation control of multi-area power system using fuzzy logic controller. *Eur. Trans. Elect. Power* 18:266–80.

67. H. Shyeghi, A. Jalili, H. A. Shayanfar. 2004. Multi-stage fuzzy load frequency control using PSO. *Energy Conversion Management* 49:377–90.

68. J. Nanda, A. Mangla. 2004. Automatic generation control of an interconnected hydro-thermal system using conventional integral and fuzzy logic controller. In *Proceedings of IEEE International Conference on Electric Utility Deregulation, Restructuring and Power Technologies*, vol. 1, pp. 372–77.

69. C. S. Indulkar, B. Raj. 2008. Application of fuzzy controller to automatic generation control. *Elect. Machine Power Syst.* 23:2570–80.

70. P. Attaviriyanupap, H. Kita, E. Tanaka, J. Hasegawa. 2004. A fuzzy-optimization approach to dynamic dispatch considering uncertainties. *IEEE Trans. Power Syst.* 19(3):1299–307.

71. M. Djukanovic, M. Novicevic, D. J. Sobajic, Y. P. Pao. 1995. Conceptual development of optimal load frequency control using artificial neural networks and fuzzy set theory. *Int. J. Eng. Intelligent Syst. Elect. Eng. Commun.* 3(2):95–108.

72. F. Beaufays, Y. Abdel-Magid, B. Widrow. 1994. Application of neural networks to load-frequency control in power systems. *Neural Networks* 7(1):183–94.

73. D. K. Chaturvedi, P. S. Satsangi, P. K. Kalra. 1999. Load frequency control: A generalised neural network approach. *Elect. Power Energy Syst.* 21:405–15.

74. H. L. Zeynelgil, A. Demirorem, N. S. Sengor. 2002. Load frequency control for power system with reheat steam turbine and governor deadband non-linearity by using neural network controller. *Eur. Trans. Elect. Power* 12(3):179–84.

75. H. Bevrani, T. Hiyama, Y. Mitani, K. Tsuji, M. Teshnehlab. 2006. Load-frequency regulation under a bilateral LFC scheme using flexible neural networks. *Eng. Intelligent Syst. J.* 14(2):109–17.

76. H. Bevrani. 2002. A novel approach for power system load frequency controller design. In *Proceedings of IEEE/PES T&D 2002 Asia Pacific*, Yokohama, Japan, vol. 1, pp. 184–89.

77. A. Demiroren, N. S. Sengor, H. L. Zeynelgil. 2001. Automatic generation control by using ANN technique. *Elect. Power Components Syst.* 29:883–96.

78. T. P. I. Ahamed, P. S. N. Rao. 2006. A neural network based automatic generation controller design through reinforcement learning. *Int. J. Emerging Elect. Power Syst.* 6(1):1–31.

79. L. D. Douglas, T. A. Green, R. A. Kramer. 1994. New approaches to the AGC nonconforming load problem. *IEEE Trans. Power Syst.* 9(2):619–28.

80. H. Shayeghi, H. A. Shayanfar. 2006. Application of ANN technique based on Mu-synthesis to load frequency control of interconnected power system. *Elect. Power Energy Syst.* 28:503–11.

81. Y. L. Abdel-Magid, M. M. Dawoud. 1996. Optimal AGC tuning with genetic algorithms. *Elect. Power Syst. Res.* 38(3):231–38.

82. A. Abdennour. 2002. Adaptive optimal gain scheduling for the load frequency control problem. *Elect. Power Components Syst.* 30(1):45–56.

83. S. K. Aditya, D. Das. 2003. Design of load frequency controllers using genetic algorithm for two area interconnected hydro power system. *Elect. Power Components Syst.* 31(1):81–94.

84. C. S. Chang, W. Fu, F. Wen. 1998. Load frequency controller using genetic algorithm based fuzzy gain scheduling of PI controller. *Elect. Machines Power Syst.* 26:39–52.

85. Z. M. Al-Hamouz, H. N. Al-Duwaish. 2000. A new load frequency variable structure controller using genetic algorithm. *Elect. Power Syst. Res.* 55:1–6.

86. A. Huddar, P. S. Kulkarni. 2008. A robust method of tuning the feedback gains of a variable structure load frequency controller using genetic algorithm optimization. *Elect. Power Components Syst.* 36:1351–68.

87. P. Bhatt, R. Roy, S. P. Ghoshal. 2010. Optimized multi area AGC simulation in restructured power systems. *Elect. Power Energy Syst.* 32:311–22.

88. A. Demirorem, S. Kent, T. Gunel. 2002. A genetic approach to the optimization of automatic generation control parameters for power systems. *Eur. Trans. Elect. Power* 12(4):275–81.

89. P. Bhatt, R. Roy, S. P. Ghoshal. 2010. GA/particle swarm intelligence based optimization of two specific varieties of controller devices applied to two-area multi-units automatic generation control. *Elect. Power Energy Syst.* 32:299–310.

90. K. Vrdoljak, N. Peric, I. Petrovic. 2010. Sliding mode based load-frequency control in power systems. *Elect. Power Syst. Res.* 80:514–27.

91. S. D. J. McArthur, E. M. Davidson, V. M. Catterson, A. L. Dimeas, N. D. Hatziargyriou, F. Ponci, T. Funabashi. 2007. Multi-agent systems for power engineering applications. Part I. Concepts, applications and technical challenges. *IEEE Trans. Power Syst.* 22(4):1743–52.

92. S. D. J. McArthur, E. M. Davidson, V. M. Catterson, A. L. Dimeas, N. D. Hatziargyriou, F. Ponci, T. Funabashi. 2007. Multi-agent systems for power engineering applications. Part II. Technologies, standards and tools for multiagent systems. *IEEE Trans. Power Syst.* 22(4):1753–59.

93. F. Daneshfar. 2009. Automatic generation control using multi-agent systems. MSc dissertation, Department of Electrical and Computer Engineering, University of Kurdistan, Sanandaj, Iran.

94. F. Daneshfar, H. Bevrani. 2010. Load-frequency control: A GA-based multi-agent reinforcement learning. *IET Gener. Transm. Distrib.* 4(1):13–26.

95. T. Hiyama, D. Zuo, T. Funabashi. 2002. Multi-agent based automatic generation control of isolated stand alone power system. Paper presented at Proceedings of International Conference on Power System Technology, Kunming, China.

96. T. Hiyama, D. Zuo, T. Funabashi. 2002. Multi-agent based control and operation of distribution system with dispersed power sources. In *Proceedings of Transmission and Distribution Conference and Exhibition*, Asia Pacific. IEEE/PES, Yokohama, Japan.

97. H. Bevrani, F. Daneshfar, P. R. Daneshmand. 2010. Intelligent power system frequency regulation concerning the integration of wind power units. In *Wind power systems: Applications of computational intelligence*, ed. L. F. Wang, C. Singh, A. Kusiak. Springer Book Series on Green Energy and Technology, 407–37. Heidelberg, Germany: Springer-Verlag.

98. H. Bevrani, F. Daneshfar, P. R. Daneshmand, T. Hiyama. 2010. Reinforcement learning based multi-agent LFC design concerning the integration of wind farms. In *Proceedings of IEEE International Conference on Control Applications*, Yokohama, Japan, CD-ROM.

99. S. P. Ghoshal. 2004. Application of GA/GA–SA based fuzzy automatic generation control of a multi-area thermal generating system. *Elect. Power Syst. Res.* 70(2):115–27.

100. S. P. Ghoshal. 2004. Optimizations of PID gains by particle swarm optimizations in fuzzy based automatic generation control. *Elect. Power Syst. Res.* 72(3):203–12.

101. S. Ganapathy, S. Velusami. 2010. MOEA based design of decentralized controllers for LFC of interconnected power systems with nonlinearities, AC-DC parallel tie-lines and SMES units. *Energy Conversion Management* 51:873–80.

102. K. Sabahi, M. Teshnehlab, M. A. Shoorhedeli. 2009. Recurrent fuzzy neural network by using feedback error learning approaches for LFC in interconnected power system. *Energy Conversion Management* 50:938–46.

103. R. Roy, P. Bhatt, S. P. Ghoshal. 2010. Evolutionary computation based three-area automatic generation control. *Expert Syst. Appl.* 37(8):5913–24.

104. T. P. I. Ahamed, P. S. N. Rao, P. S. Sastry. 2002. A reinforcement learning approach to automatic generation control. *Elect. Power Syst. Res.* 63:9–26.

105. Y. L. Karnavas, D. P. Papadopoulos. 2002. AGC for autonomous power system using combined intelligent techniques. *Elect. Power Syst. Res.* 62:225–39.

106. J. Nanda, S. Mishra, L. C. Saikia. 2009. Maiden application of bacterial foraging-based optimization technique in multiarea automatic generation control. *IEEE Trans. Power Syst.* 24(2):602–9.

107. R. D. Chritie, A. Bose. 1996. Load frequency control issues in power system operation after deregulation. *IEEE Trans. Power Syst.* 11(3):1191–200.

108. J. Kumar, N. G. K. Hoe, G. B. Sheble. 1997. AGC simulator for price-based operation. Part I. A model. *IEEE Trans. Power Syst.* 2(12):527–32.
109. J. Kumar, N. G. K. Hoe, G. B. Sheble. 1997. AGC simulator for price-based operation. Part II. Case study results. *IEEE Trans. Power Syst.* 2(12):533–38.
110. V. Donde, M. A. Pai, I. A. Hiskens. 2001. Simulation and optimization in a AGC system after deregulation. *IEEE Trans. Power Syst.* 16(3):481–89.
111. B. Delfino, F. Fornari, S. Massucco. 2002. Load-frequency control and inadvertent interchange evaluation in restructured power systems. *IEE Proc. Gener. Transm. Distrib.* 149(5):607–14.
112. H. Bevrani, Y. Mitani, K. Tsuji. 2004. Robust AGC: Traditional structure versus restructured scheme. *IEEJ Trans. Power Energy* 124-B(5):751–61.
113. G. Dellolio, M. Sforna, C. Bruno, M. Pozzi. 2005. A pluralistic LFC scheme for online resolution of power congestions between market zones. *IEEE Trans. Power Syst.* 20(4):2070–77.
114. H. Bevrani. 2004. Decentralized robust load-frequency control synthesis in restructured power systems. PhD dissertation, Osaka University.
115. B. H. Bakken, O. S. Grande. 1998. Automatic generation control in a deregulated power system. *IEEE Trans. Power Syst.* 13(4):1401–6.
116. B. Tyagi, S. C. Srivastava. 2006. A decentralized automatic generation control scheme for competitive electricity market. *IEEE Trans. Power Syst.* 21(1):312–20.
117. J. M. Arroyo, A. J. Conejo. 2002. Optimal response of a power generator to energy, AGC, and reserve pool-based markets. *IEEE Trans. Power Syst.* 17(2):404–10.
118. F. Liu, Y. H. Song, J. Ma, S. Mei, Q. Lu. 2003. Optimal load-frequency control in restructured power systems. *IEE Proc. Gener. Transm. Distrib.* 150(1):377–86.
119. S. Bhowmik, K. Tomsovic, A. Bose. 2004. Communication models for third party load frequency control. *IEEE Trans. Power Syst.* 19(1):543–48.
120. H. Bevrani, Y. Mitani, K. Tsuji, H. Bevrani. 2005. Bilateral-based robust load-frequency control. *Energy Conversion Management* 46:1129–46.
121. H. Bevrani, T. Hiyama. 2007. Robust decentralized PI based LFC design for time-delay power systems. *Energy Conversion Management* 49:193–204.
122. H. Outhred, S. R. Bull, S. Kelly. 2007. Meeting the challenges of integrating renewable energy into competitive electricity industries. http://www.reilproject.org/documents/GridIntegrationFINAL.pdf.
123. Department of Trade and Industry. 2006. *The energy challenge energy review report.* London: DTI.
124. EWIS. 2007. *Towards a successful integration of wind power into European electricity grids.* Final report. http://www.ornl.gov/~webworks/cppr/y2001/rpt/122302.pdf.
125. AWEA Resources. 2008. U.S. wind energy projects. The American Wind Energy Association. http://www.awea.org.
126. M. Yamamoto, O. Ikki. 2010. National survey report of PV power applications in Japan 2009. Int. Energy Agency. Available: http://www.iea-pvps.org/countries/download/nsr09/NSR_2009_Japan_100620.pdf.
127. The Global Wind Energy Council. 2008. US, China & Spain lead world wind power market in 2007. *GWEC Latest News.* http://www.gwec.net/ (accessed February 28, 2008).
128. NERC Special Report. 2009. Accommodating high levels of variable generation. http://www.nerc.com/files/IVGTF_Report_041609.pdf (accessed May 17, 2010).

129. H. Bevrani, A. Ghosh, G. Ledwich. 2010. Renewable energy resources and frequency regulation: Survey and new perspectives. *IET Renewable Power Gener.*, 4(5): 438–57.

130. S. Nomura, Y. Ohata, T. Hagita, et al. 2005. Wind farms linked by SMES systems. *IEEE Trans. Appl. Superconductivity* 15(2):1951–54.

131. G. Strbac, A. Shakoor, M. Black, et al. 2007. Impact of wind generation on the operation and development of the UK electricity systems. *Elect. Power Syst. Res.* 77:1214–27.

132. H. Banakar, C. Luo, B. T. Ooi. 2008. Impacts of wind power minute to minute variation on power system operation. *IEEE Trans. Power Syst.* 23(1):150–60.

133. G. Lalor, A. Mullane, M. O'Malley. 2005. Frequency control and wind turbine technology. *IEEE Trans. Power Syst.* 20(4):1905–13.

134. J. Morren, S. W. H. de Haan, W. L. Kling, et al. 2006. Wind turbine emulating inertia and supporting primary frequency control. *IEEE Trans. Power Syst.* 21(1):433–34.

135. C. Luo, H. Golestani Far, H. Banakar, et al. 2007. Estimation of wind penetration as limited by frequency deviation. *IEEE Trans. Energy Conversion* 22(2):783–91.

136. P. Rosas. 2003. Dynamic influences of wind power on the power system. PhD dissertation, Technical University of Denmark.

137. P. R. Daneshmand. 2010. Power system frequency control in the presence of wind turbines. Master's thesis, Department of Computer and Electrical Engineering, University of Kurdistan.

138. N. R. Ullah, T. Thiringer, D. Karlsson. 2008. Temporary primary frequency control support by variable speed wind turbines: Potential and applications. *IEEE Trans. Power Syst.* 23(2):601–12.

139. J. L. R. Amenedo, S. Arnalte, J. C. Burgos. 2002. Automatic generation control of a wind farm with variable speed wind turbines. *IEEE Trans. Energy Conversion* 17(2):279–84.

140. E. Hirst. 2002. Integrating wind output with bulk power operations and wholesale electricity markets. *Wind Energy* 5(1):19–36.

141. R. Doherty, H. Outhred, M. O'Malley. 2006. Establishing the role that wind generation may have in future generation portfolios. *IEEE Trans. Power Syst.* 21:1415–22.

142. H. Holttinen. 2005. Impact of hourly wind power variation on the system operation in the Nordic countries. *Wind Energy* 8(2):197–218.

143. H. Bevrani, A. G. Tikdari. 2010. An ANN-based power system emergency control scheme in the presence of high wind power penetration. In *Wind power systems: Applications of computational intelligence*, ed. L. F. Wang, C. Singh, A. Kusiak, 215–54. Springer Book Series on Green Energy and Technology. Heidelberg: Springer-Verlag.

144. J. G. Slootweg, W. L. Kling. 2003. The impact of large scale wind power generation on power system oscillations. *Elect. Power Syst. Res.* 67:9–20.

145. C. Chompoo-inwai, W. Lee, P. Fuangfoo, et al. 2005. System impact study for the interconnection of wind generation and utility system. *IEEE Trans. Industry Appl.* 41:163–68.

146. SmartGrids. www.smartgrids.eu (accessed May 17, 2010).

147. IntelliGrid architecture. www.intelligrid.info/intelliGrid_architecture/Overview_Guidelines/index.htm (accessed May 17, 2010).

148. Gridwise Alliance. www.gridwise.org/ (accessed May 17, 2010).

149. A. Mehrizi-Sani, R. Iravani. 2009. Secondary control for microgrids using potential functions: Modelling issues. Paper presented at Proceedings of CIGRE Conference on Power Systems, Toronto.
150. G. Diaz, C. Gonzalez-Moran, J. Gomez-Aleixandre, A. Diez. 2010. Scheduling of droop coefficients for frequency and voltage regulation in isolated microgrids. *IEEE Trans. Power Syst.* 25(1):489–96.
151. M. C. Chandorkar, D. M. Divan, R. Adapa. 2007. Control of parallel connected inverters in standalone ac supply-systems. *IEEE Trans. Ind. Appl.* 22(4):136–43.
152. M. Adamiak, S. Bose, Y. Liu, K. Bahei-Eldin, J. deBedout. Tieline controls in microgrid applications. http://www.gedigitalenergy.com/smartgrid/May08/4_Microgrid_Applications.pdf (accessed May 17, 2010).
153. K. Fujimoto, et al. 2009. Load frequency control using storage system for a micro grid. Paper presented at Proceedings of IEEE T&D Asia Conference and Exposition, Seoul.
154. M. G. Molina, P. E. Mercado. 2009. Control of tie-line power flow of microgrid including wind generation by DSTATCOM-SMES controller. Paper presented at Proceedings of IEEE Energy Conversion Congress and Expo (ECCE) San Jose, CA.
155. T. Chaiyatham, I. Ngmroo, S. Pothiya, S. Vachirasricirikul. 2009. Design of optimal fuzzy logic-PID controller using bee colony optimization for frequency control in an isolated wind-diesel system. Paper presented at Proceedings of IEEE T&D Asia Conference and Exposition, Seoul.
156. A. Madureia, C. Moreira, J. Pecas Lopes. 2005. Secondary load-frequency control for microgrids in islanded operation. Paper presented at Proceedings of International Conference on Renewable Energy and Power Quality (ICREPQ05), Zavagoza, Spain.
157. N. J. Gil, J. A. Pecas Lopes. 2007. Hierarchical frequency control scheme for islanded multi-microgrids operation. Paper presented at Proceedings of IEEE Power Tech, Lausanne, Switzerland.
158. T. Hiyama, et al. 2004. Multi-agent based operation and control of isolated power system with dispersed power sources including new energy storage device. In *Proceedings of International Conference on Renewable Energies and Power Quality (ICREPQ'04)*, CD-ROM.
159. X. Li, Y. J. Song, S. B. Han. 2008. Frequency control in micro-grid power system combined with electrolyzer system and fuzzy PI controller. *J. Power Sources* 180:468–75.
160. J. M. Manuel, et al. 2009. Frequency regulation contribution through variable-speed wind energy conversion systems. *IEEE Trans. Power Syst.* 24(1):173–80.
161. National Laboratory for Sustainable Energy. 2009. *The intelligent energy system infrastructure for the future*, ed. H. Larsen, L. S. Petersen. RisØ Energy Report, vol. 8. National Laboratory for Sustainable Energy, Roskilde, Denmark.
162. M. Tsili, S. Papathanassiou. 2009. A review of grid code technical requirements for wind farms. *IET Renew. Power Gener.* 3(3):308–32.

4

AGC in Restructured Power Systems

As the world moves toward competitive markets in electric power systems, the shift of ownership and operational control of generation from the vertically integrated utilities to independent, for-profit generation owners has raised a number of fundamental questions regarding AGC systems. Key questions relate to the new AGC designs that are more appropriate to the new operational objectives of a restructured power network, including the revising of traditional control schemes and the AGC model by taking into account bilateral transactions.

After deregulation of the electricity sector, all reliability entities in the world, such as the North American Electric Reliability Council (NERC) and Union for the Coordination of Transmission of Electricity (UCTE), updated the control performance standards for AGC. The crucial role of AGC systems will continue in restructured power systems, with some modifications to account for some issues, such as bilateral contracts and deregulation policy among the control areas. In a real-time power market, AGC as an ancillary service provides an essential role for ensuring reliable operation by adjusting generation to minimize frequency deviations and regulate tie-line flows.

This chapter reviews the main structures, configurations, and characteristics of AGC systems in a deregulated environment. Section 4.1 addresses the control area concept in restructured power systems. Modern AGC structures and topologies are described in Section 4.2. A brief description of AGC markets is addressed in Section 4.3. Some concepts of the AGC market and market operator, the needs for intelligent AGC markets in the future, and also an updated conventional frequency response model concerning the bilateral transactions are explained in Section 4.4. Finally, the chapter is concluded in Section 4.5.

4.1 Control Area in New Environment

Most deregulated utilities have chosen to control the frequency and tie-line power to the same quality as before deregulation. The AGC schemes and control strategies have also mostly remained similar to before deregulation, except that some definitions are changed and services provided by participants are now classified as ancillary. However, the introduction of electricity markets has added to the pressures to redefine some concepts and to update the way that frequency/real power is controlled.

In an open energy market, generation companies (Gencos) as independent power utilities may or may not participate in the AGC task. On the other hand, distribution companies (Discos) may contract individually with Gencos, renewable energy plants, or independent power producers (IPPs) for power in different areas. Therefore, in the new environment, control is highly decentralized. Each load matching contract requires a separate control process, yet this process must cooperatively interact to reestablish system frequency and tie-line power interchange.[1] In real-time markets, new organizations, market operators, and supervisors, such as independent system operators (ISOs), are responsible for maintaining the real-time balance of generation and load for minimizing frequency deviations and regulating tie-line flows, which would facilitate bilateral contracts spanning over various control areas.

In the new structure, there are no constant boundaries for control areas. The definition of a *control area* is somewhat determined by pooling arrangements and contract agreements of AGC participating utilities. The boundary of the control area encloses the Gencos, transmission company (Transco), and Discos associated with the performed contracts. In order to supply the load, the Discos can get power from Gencos directly or through Transco. Such a configuration is conceptually shown in Figure 4.1. The control areas are interconnected to each other, through either the Transco or Gencos.[2]

In a modern power system, the AGC system should track moment-to-moment fluctuations in the system load to meet the specified control area performance criteria, such as those criteria provided by NERC and UCTE.

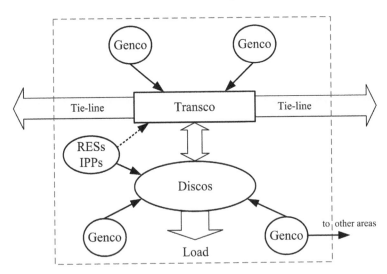

FIGURE 4.1
A (virtual) control area in a deregulated environment.

These criteria require the AGC system to maintain the area control error (ACE) within tight limits. All control areas in a multiarea power system are required to follow the determined control performance standards. The assigned control area performance criteria are measurable and in use for normal functions of each control area's energy management system (EMS).

Currently, in many countries, electric systems are restructured; new market concepts were adopted to achieve the goal of better performance and efficiency. Operating the power system in a new environment will certainly be more complex than in the past, due to restructuring and a considerable degree of technical and economical interconnections. In addition to various market policies, numerous generator units in distribution areas and a growing number of independent players and renewable energy sources (RESs) are likely to impact on the operation and control of the power system.

The classical AGC scheme may not be as straightforward to use in a deregulated power system, which includes separate Gencos, RESs, IPPs, Discos, and Transcos in a competitive environment, as for vertically integrated utility structures. In response to the new challenges, novel modeling and control approaches are required to maintain reliability, to follow AGC tasks, and to get a new trade-off between efficiency and robustness.[1] As the electric power industry moves toward full competition, various industry consensus on definitions, requirements, obligations, and management for AGC ancillary service is being developed by many entities across the world, such as NERC, UCTE, the Federal Energy Regulatory Commission (FERC), Oak Ridge National Laboratory (ORNL), etc.

4.2 AGC Configurations and Frameworks

4.2.1 AGC Configurations

The MW-frequency regulation issue in a multiarea power system is mainly referred to as frequency control, load following, and scheduling. The main difference between frequency control and load following issues is in the timescale over which these fluctuations occur. Frequency control responds to rapid load fluctuations (on the order of a few seconds to a minute), and load following responds to slower changes (on the order of a few minutes). While frequency regulation matches the generation with a seconds-to-minutes load change, load following uses the generation to meet minutes-to-hour and daily variations of load.[3] These issues are addressed by governor systems, AGC, and economic dispatch mechanisms.[4]

In practice, AGC configurations could differ according to their timing, the amount of information individual suppliers and loads provide to the market operator, and the role of the market operator in facilitating or directing this ancillary service. A general configuration for the AGC system in a deregulated environment is shown in Figure 4.2. The Gencos send the bid regulating reserves to the AGC center through a secure network service. These bids are sorted by a prespecified time period and price. Then, the sorted regulating reserves with the demanded load from Discos, the tie-line data from Transco, and the area frequency are used to provide control commands to track the area load changes. The bids are checked and re-sorted according to the received congestion information (from Transco) and screening of available capacity (collected from Gencos). The control signal is transmitted to the Gencos once every one to few seconds, while the results of computing participation factors and load generation scheduling by the market operator (economic dispatch unit in the AGC center) are executed daily or every few hours.

A general scheme for AGC participants in restructured power systems can be considered as shown in Figure 4.3. The Gencos (and many distributed power producers) would interact with the market operator by providing bids for the supporting AGC service. In fact, the market operator is responsible for

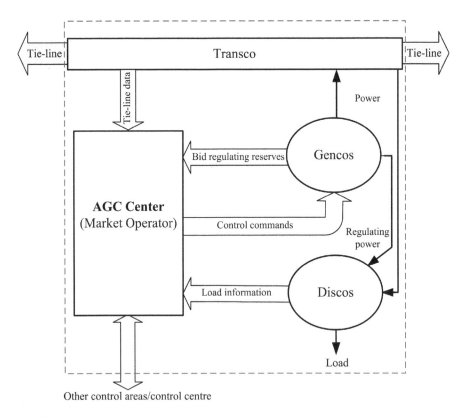

FIGURE 4.2
General AGC configuration in a deregulated environment.

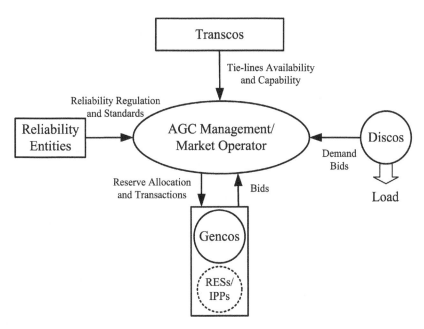

FIGURE 4.3
AGC participants in a deregulated environment.

trading the regulation power. Transcos also post their information regarding the availability and capability of transmission lines via a secure communication system. Discos submit their demand bids to the market operator to be matched with Genco's bids while satisfying the performed regulation standards provided by reliability entities.

In many market models, such as Poolco,[5,6] since the AGC ancillary service auction is operated by the market operator, the market operator is the single-buyer party to meet the reliability obligations. The main market objective is to minimize AGC payments to Gencos while encouraging Gencos to provide sufficient regulation power. Gencos would anticipate submitting a bid that would maximize their profits as allocations are made. The AGC bids should include financial information for capacity reservation and energy, as well as operational information, such as location, ramp rate, and quantity blocks. Based on the central operator requirements for the AGC issue and participants' bids submitted to the market operator, the price and quantity of regulation power are determined, and payments are calculated by the market operator.

The above-mentioned mechanism mostly deals with a *centralized* AGC market (usually called pool market), which is cleared by a unique market operator that collects the offers to sell and the bids to buy. On the other hand, in a *decentralized* AGC market (usually called exchanges market), sellers and buyers can enter directly into contracts to buy and sell. In this case, the transaction is of a bilateral type.

As mentioned, with a centralized structure, the market is cleared by a unique entity. In pool-energy-only markets, which are popular across the United States, participants provide bids and offers. Then, the system operator directly commits and dispatches the producers. Therefore, this approach favors links between AGC services and other products, such as energy. However, markets with a centralized structure are deemed to be opaque because the clearing process is quite complex. In addition, bidders have to provide a lot of information, and this system hardly takes into account all the variables of the system. In a real-time power market, the energy component of an ancillary services bid is used for energy balancing and ex-post pricing systems. Resources available in the energy balancing system include regulation, spinning, nonspinning, and replacement reserves, as well as resources for submitted supplemental bids for real-time imbalances that are pooled in the energy balancing system and arranged in merit order based on their energy bid prices.[7] In the decentralized structure, which is popular across Europe, participants propose bids and select offers directly in the market. Therefore, a global co-optimization is difficult since participants buy and sell independently from each other. Instead, each participant does its own co-optimization with its assets.[8]

In defining an AGC market, a key factor is to attract enough regulation power producers to make the market competitive, while maintaining an acceptable level of security and reliability. If too much energy is traded close to real time, then the market operator must contract more reserves to ensure that the predicted demand can be met. One of the principles recommended in the FERC standard market design is that the design should ensure there are few incentives for a participant not to be in balance prior to real time.[9] The AGC performance can also be significantly influenced by the deregulation policy,[5,7,10] communication structure/facilities,[11,12] and most importantly, the reserve levels.[13]

The market operator usually provides a priority list sorting by regulating price as described in Bevrani.[1] The capacity of one regulating object in the table should be fully utilized before calling on the next, which is more expensive. This applies to both cost and the market environment. In essence, this would mean that at any given time the cheapest solution should be in place. However, this is often far from the reality, and due to many reasons, the economic solution is not what it should be. Reserve levels also need to be considered, as a cheap generator might have its output reduced to ensure sufficient reserve levels. The AGC algorithm needs to be set up so that an expensive generator decreases and a cheap generator increases its regulation power, simultaneously.[14]

4.2.2 AGC Frameworks

In the real world, different AGC frameworks/schemes are available to perform supplementary control among different countries/regions. The AGC scheme that has been implemented in some countries differs from the design adopted in most other parts of the world. Considerable differences exist between the AGC characteristics, reserve service, topology, and related standards defined in various jurisdictions. This diversity is the source of some confusion because the diversity is extended not only to the specification of existing AGC systems, but also to the terms used to describe them. Below, the general AGC framework provided by UCTE is briefly explained. In the UCTE terminology, instead of AGC, the secondary control or load-frequency control (LFC) term is used.

According to the UCTE definitions,[15] a *control area* is the smallest portion of a power system equipped with an autonomous AGC system. A *control block* may be formed by one or more control areas working together to satisfy predefined AGC performance requirements with respect to the neighboring control blocks, within a *synchronous area* connected to the UCTE network.

Within the synchronous area, the control actions and reserves are organized in a hierarchical structure with control areas, and control blocks with a coordination center. The AGC, the technical reserves, and the corresponding control performances are essential to allow transmission system operators (TSOs) or block coordinators to perform daily operational business.

As shown in Figure 4.4, the synchronous area consists of multiple interconnected control areas/blocks, each of them with a centralized supplementary control loop. Each control area/block may divide up into subcontrol areas that operate their own underlying AGC, as long as this does not jeopardize the interconnected operation. Figure 4.4 shows the hierarchy of an AGC that consists of the synchronous area, with control blocks and (optionally) included control areas.

If a control block has internal control areas, the control block organizes the internal frequency regulation according to one of the centralized, pluralistic, and hierarchical schemes.[15] In a *centralized* scheme, the AGC for the control block is performed centrally by a single controller (the control block

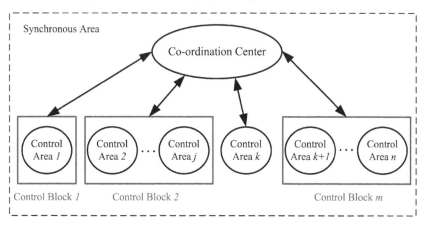

FIGURE 4.4
A UCTE synchronous area.

including only one control area); the operator of the control block has the same responsibilities as the operator of a control area. This scheme is currently in use in some UCTE countries, such as Italy, Austria, Belgium, and the Netherlands.

In a *pluralistic* scheme, the AGC is performed in a decentralized way with more than one control area; a single TSO (block coordinator) controls the whole block toward its neighbors, with its own supplementary loop and regulating capacity, while all the other TSOs of the block regulate their own control areas in a decentralized way by their own. This scheme exists in France, Spain, and Portugal, with France operating as the block coordinator.

Finally, in a *hierarchical* scheme the AGC is performed in a decentralized way with more than one control area. The AGC is carried out by several separate supplementary control loops, one for each control area within the hierarchical control block. They are separately controlling their cross-border exchanges. But at a higher control level, a single TSO (block coordinator) operates the superposed block controller that directly influences the subordinate supplementary loops of all control areas of the control block; the block coordinator may or may not have regulating capacity on its own. Switzerland represents an application of this AGC scheme.

In particular, most European power systems in UCTE use one of the aforementioned control schemes. However, there are important differences among them in details, and some AGC schemes exhibit some important differences with respect to the standard structures.[16-20]

4.3 AGC Markets

After the advent of deregulation, there was much effort to form competitive markets for ancillary services. Currently in many countries, similar to available competitive markets, some markets exist for ancillary services, such as the *AGC market*, with different structures and even titles, depending on the rules and regulations of the area. However, since a Genco can make the choice to allocate 1 MW of production capacity as energy or an AGC service, markets for AGC services and energy are closely linked. Furthermore, AGC markets as well as energy markets are highly influenced by other markets and commodities, such as fuel and environment markets.

Systems across the world have adopted different methods to calculate the needs for regulation services, which leads to different types of AGC markets. In North America, regulation reserve markets for AGC with fully dispatchable regulation power capacity within 10 min are available. In some regions, such as England (without an official AGC system), AGC market is summarized to a part of the spinning reserve as a power exchange system. It provides a system with a 30 min short-term market for balancing, operating

1 h ahead of real time. However, in many countries, such as Japan, Australia, China, the Nordic countries, and continental Europe, the AGC markets and frequency performance standards are strongly influenced by grid rules and regional policies.[21]

Flat rate, price based, and response based are three famous AGC markets.[22] *Flat rate*, because of simplicity, is the most common type of AGC market (specifically in North America and North Africa). It provides a 10 min regulation market with a uniform price payment at the rate of the market clearing price (MCP), without considering the ramp rates and response quality of the participant generating units. A *price-based* AGC market provides a 5 or 10 min regulation market, and the participant generating units are paid based on their ramp rate performance. Finally, the *price-based* AGC market provides two separate markets for fast ramp regulation (5 min market) and slow ramp regulation (10 min market), and the AGC participant generating units are paid at the rate of MCP as the maximum possible payment available in each auction market.

It has been demonstrated that in comparison to the flat-rate AGC market, the price-based and ramp-rate-based AGC markets increase competition, encourage participant generators with incentive, and explore more control options to optimize AGC performance. The separation of fast and slow ramp generators in the response-based AGC market makes it possible for this market to call upon the appropriate service, depending on the magnitude of the disturbance that is suitable for contingencies and related power reserves. However, in this case (separate markets for fast ramp and slow ramp regulation), making a decision is more difficult. To procure a certain amount of regulation from such a market, the market operator has to decide how much of the fast and slow regulation powers are to be bought. Then there are multiple options available as to how to use them in time of need.[22]

All the generators participating in the AGC markets mentioned above are required to meet specified technical and operating requirements, and also, they should determine regulation capacity, price, and operational ramp rate (MW/min) in their bids. The AGC markets are usually cleared for every dispatch interval during the trading interval ahead of real time.[23]

In the AGC markets, the structures of bids are related to the scoring and clearing processes, while the structures of payments are related to the settlement process. In the literature, most of the discussions on structures of offers and payment of AGC services are concentrated on *capacity*, *utilization*, and *opportunity cost* components.[24–27] The more common structures for offers and payments are known as (1) a fixed allowance, (2) an availability price, (3) a price for kinetic energy, (4) a utilization payment, (5) a utilization frequency payment, and (6) a payment for the opportunity cost.[8] A fixed allowance is paid to the provider in every instance. An availability price is paid only when the unit is in a ready-to-provide state. A price for kinetic energy remunerates the quantity of kinetic energy made available to the system. It recognizes the machines with high kinetic energy, and thus high inertia to shape the rate of

frequency change following a contingency. A utilization payment remunerates the actual delivery of the service. A utilization frequency payment is
based on the number of calls to provide a service over a given period of time.
It thus reflects the extra costs that may be incurred each time the service is
called upon. The payment of the opportunity cost has been identified for a
long time by the community as an important allowance.

Coordination of the AGC market with other ancillary and energy markets
is also an important problem. Since a generating unit may provide several
ancillary services (including AGC), and thus contribute to several markets,
coordination between different markets, in both quantity and price issues,
is required. For example,[8] if a generator provides reserves for an AGC system, it cannot sell all its capacity on the only energy market. On the other
hand, if a generator is committed through energy dispatch, it is then able to
provide reactive power support for voltage control service. A direct consequence of this feature is that the prices of ancillary services and electrical
energy will interact.

4.4 AGC Response and an Updated Model

4.4.1 AGC System and Market Operator

As mentioned, the main objectives of an AGC system are to maintain the frequency within control areas close to the nominal value, as well as to control
tie-line flows at scheduled values defined by utilities' contracts. Similar to
the conventional AGC system, the balance between generation and load can
be achieved by detecting frequency and tie-line flow deviations via ACE signal through an integral feedback control mechanism. If supply and demand
do not match in the long run, as well as in the short run, the market will fail.
The supply of AGC services is mostly ensured by conventional generating
units. Marginally, other participants also provide regulation services, such
as storage devices that smooth either consumption or generation, consumers
that can modulate their consumption upon request or automatically, and to
some extent, RESs. The demand for AGC services is defined by the market
operator and depends on the power system structure.

As explained in Chapter 2, the generating units could respond to fast load
fluctuations, on a timescale of 1 to 3 s, depending on the droop characteristics of governors in the primary frequency control loop. The generating
units could respond to slower disturbance dynamics in the range of a few
seconds, measuring the ACE signal via a supplementary frequency control
loop in the AGC system. The longer-term load changes on a timescale of 10
s to several minutes could respond based on economic dispatch plans and
special control actions that would utilize the economics of the AGC system
to minimize operating costs.

The deviations in load and power could be procured by the market operator on purpose, because of planned line and unit outages. This kind of deviation may be produced by the market operator as a control plan in response to energy imbalances following unpredicted disturbances. These deviations are basically different than unpredicted frequency/tie-line deviations that usually occur by variations of load and generation from scheduled levels following a fault, such as unplanned line and unit outages. The AGC participant generating units in an AGC market could respond to unpredicted frequency/tie-line deviations proportional to the assigned *participation factors* from their schedules within a few seconds. The market operator will change the set points of AGC units, which have submitted energy price/quantity bids for the real-time energy imbalances, by means of a new control plan, as shown in Figure 4.5.

Determining AGC participation factors by market operator is an important issue in deregulated environments. For this purpose, several factors, such as regulation price, ramp rate, and bid capacity provided by the candidate generating units, should be considered. The impacts of these factors on AGC performance and system frequency response characteristics, including maximum frequency deviation, the time taken to bring the frequency back within safe limits, and the time taken by ACE to cross zero for the first time following the disturbance, are studied in PSERC.[22] It has been shown that the frequency response is better when the participation factors are proportional to the units' ramp rate.

The market operator may procure the required power regulation from various existing reserves, such as normal AGC regulation, spinning reserves that are usually available within 10 min, nonspinning reserves, and replacement reserves. In this process, the participant Gencos would be allowed to rebid their uncommitted resources and regulation powers at new prices. The market operator, which is responsible for AGC procurement, can use various methods to obtain AGC services. Some methods are known as compulsory

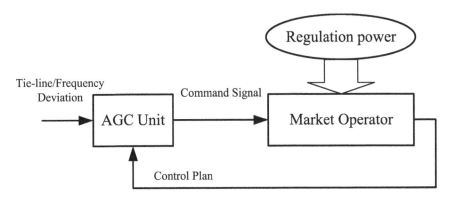

FIGURE 4.5
The AGC–market operator loop.

provision, bilateral contracts, tendering, self-procurement, and spot market. These methods are defined in Rebours.[8] Various factors, such as market concentration, mode of energy and transmission trading, risk aversion, the costs recovery method, and centralized or decentralized AGC control, influence the choice of one of these methods over the others.

In addition to the quality and quantity of AGC services, the location is also important. Although frequency control ancillary services act on global frequency, their physical locations should be considered while procuring AGC services for some reason, mostly with the security and reliability issues.[8] Congestion of transmission lines is an important reason that can affect the reliable provision of AGC services. If enough transmission capacity is not available, the affected zone has to secure enough ancillary services from within its perimeter. Therefore, a part of the transmission capacity has to be allocated for AGC services. This congestion of transmission lines and overloading is more important in the presence of contingencies and emergency conditions, and it should be carefully considered for performing emergency control actions.[28,29]

Regarding the high cost of reserving transmission capacity, the contributions to the AGC are likely to be distributed across the whole power system network to reduce unplanned power transits following a large generation outage.[8] A distributed framework for AGC services can also be useful following the islanding issue. Islands cannot stay stable without any frequency control system service.

In fact, trading AGC services in a distributed framework across systems allows more efficient use of flexible resources, reduces the potential exercise of market power, diminishes imbalance exposure, and makes better use of interconnection capabilities.[8,30] A necessary condition to perform such a useful framework is using a distributed generation scheme across the whole interconnected network.

In a competitive environment with a decentralized market structure, a Disco has the freedom to contract with any Genco in its own area or sign bilateral contracts with a Genco in another area that would be cleared by the market operator. If a bilateral contract exists between Discos in one control area and Gencos in other control areas, the scheduled flow on a tie-line between two control areas must exactly match the net sum of the contracts that exist between market participants on opposite sides of the tie-line. If the bilateral contract is adjusted, the scheduled tie-line flow must be adjusted accordingly. In general, using bilateral contracts, Discos would correspond demands to Gencos, which would introduce new signals that did not exist in the vertically integrated environment. These signals would give information as to which Genco ought to follow which Disco. Moreover, these signals would provide information on scheduled tie-line flow adjustments and ACEs for control areas.[7]

In a competitive electricity environment, Poolco and bilateral transactions may take place simultaneously. As already mentioned, in Poolco-based transactions, the power generating units and consumers submit their bids to the market operator, and market players quote a price and quantity for upward and downward adjustment. For each time period of operation, generators' bids are selected based on the principle of the cheapest bid first for upward regulation and the most expensive bid first for the downward regulation. In addition, during the low-frequency conditions, consumers having interruptible loads may also be selected based on the cheapest bid first. The resultant price/quantity list is used for achieving the balance between consumption and production.

The AGC in a deregulated electricity market should be designed to consider different types of possible transactions,[5,31,32] such as Poolco-based transactions, bilateral transactions, and their combination. Over the last years, some published works have addressed the updating of the traditional AGC model and the redesign of conventional control schemes to accommodate bilateral transactions.[31–34] An AGC scheme required for Poolco-based transactions, utilizing an integral controller, has been suggested in the literature.[5,31,32,35]

4.4.2 AGC Model and Bilateral Contracts

Using the idea presented in Donde et al.[31] the well-known AGC frequency response model (Figure 2.11) can be updated for a given control area in a deregulated environment with bilateral transactions. The result is shown in Figure 4.6. This model uses all the information required in a vertically operated utility industry plus the contract data information.

The overall power system structure can be considered as a collection of Discos or control areas interconnected through high-voltage transmission lines or tie-lines. Each control area has its own AGC and is responsible for tracking its own load and honoring tie-line power exchange contracts with its neighbors. There can be various combinations of contracts between each Disco and available Gencos. On the other hand, each Genco can contract with various Discos. Therefore, a Disco in any of the areas and Gencos in the same or in a different area may also negotiate bilateral contracts. These players of the electricity market are responsible for having a communication path to exchange contract data, as well as measurements to perform the load following function. In such contracts, a Genco changes its power output to follow the predicted load as long as it does not exceed the contracted value.

The generation participation matrix (GPM) concept is defined to express these bilateral contracts in the generalized model.[1] GPM shows the participation factor of each Genco in the considered control areas, and each control area is determined by a Disco. The rows of a GPM correspond to Gencos, and columns to control areas that contract power. For example, for a large-scale power system with m control areas (Discos) and n Gencos, the GPM will have the following structure, where gpf_{ij} refers to the generation participation

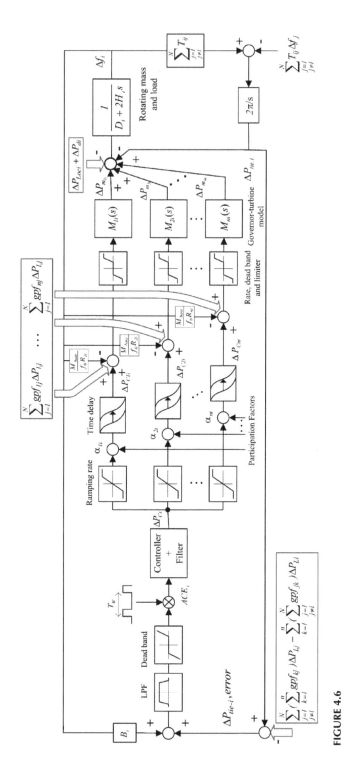

FIGURE 4.6
An updated AGC response model for deregulated environments.

factor and shows the participation factor of Genco i in the load following of area j (based on a specified bilateral contract):

$$GPM = \begin{bmatrix} gpf_{11} & gpf_{12} & \cdots & gpf_{1(m-1)} & gpf_{1m} \\ gpf_{21} & gpf_{22} & \cdots & gpf_{2(m-1)} & gpf_{2m} \\ \vdots & \vdots & \vdots & \vdots & \vdots \\ gpf_{(n-1)1} & gpf_{(n-1)2} & \cdots & gpf_{(n-1)(m-1)} & gpf_{(n-1)m} \\ gpf_{n1} & gpf_{n2} & \cdots & gpf_{n(m-1)} & gpf_{nm} \end{bmatrix} \tag{4.1}$$

New information signals due to possible various contracts between Disco i and other Discos and Gencos are shown as wide arrows. Here, we can write:

$$\Delta P_{tie-i,\,error} = \Delta P_{tie-i,\,actual} - \sum (\text{Total export power} - \text{Total import power})$$

$$= \Delta P_{tie-i,\,actual} - \sum_{\substack{j=1 \\ j\neq i}}^{N} \left(\sum_{k=1}^{n} gpf_{kj} \right) \Delta P_{Lj} - \sum_{k=1}^{n} \left(\sum_{\substack{j=1 \\ j\neq i}}^{N} gpf_{jk} \right) \Delta P_{Li} \tag{4.2}$$

where

$$\sum_{i=1}^{n} gpf_{ij} = 1, \quad \sum_{k=1}^{n} \alpha_{ki} = 1; \qquad 0 \leq \alpha_{ki} \leq 1 \tag{4.3}$$

$$\Delta P_{mi} = \sum_{j=1}^{N} gpf_{ij} \Delta P_{Lj} \tag{4.4}$$

where ΔP_{di} (in Figure 4.6) is the area load disturbance, ΔP_{Loc-i} is the contracted load demand (contracted and uncontracted) in area i, and $\Delta P_{tie-i,\,actual}$ is the actual tie-line power in area i. Using Equation 4.2, the scheduled tie-line power ($\Delta P_{tie-i,\,scheduled}$) can be calculated as follows:

$$\Delta P_{tie-i,\,scheduled} = \sum_{\substack{j=1 \\ j\neq i}}^{N} \left(\sum_{k=1}^{n} gpf_{kj} \right) \Delta P_{Lj} - \sum_{k=1}^{n} \left(\sum_{\substack{j=1 \\ j\neq i}}^{N} gpf_{jk} \right) \Delta P_{Li} \tag{4.5}$$

Interested readers can find more details and simulation results on the above generalized AGC scheme for restructured power systems in the literature.[1,32,36]

4.4.3 Need for Intelligent AGC Markets

The AGC markets of tomorrow, which should handle complex multiobjective regulation optimization problems characterized by a high degree of diversification in policies, control strategies, and wide distribution in demand and supply sources, must be intelligent. The core of such an intelligent system should be based on flexible intelligent algorithms, advanced information technology (IT), and fast communication devices.

The intelligent AGC market interacting with ancillary services and energy markets will be able to contribute to upcoming challenges of future power systems control and operation. This issue will be performed by intelligent meters and data analyzers using advanced computational methods and hardware technologies in both load and generation sides. The future AGC requires increased intelligence and flexibility to ensure that they are capable of maintaining a supply-load balance following serious disturbances.

4.5 Summary

During the last two decades, energy regulatory policies all around the world have been characterized by the introduction of competition in many electric power systems. The AGC issue as an ancillary service represents an important role to maintain an acceptable level of efficiency, quality, and reliability in a deregulated power system environment. To this aim, researchers and responsible organizations have started to analyze possible new AGC schemes and regulation solutions, with paradigms suited for the energy market scenarios. These new solutions can rely on recent advances in IT, artificial intelligent methodologies, and innovations in control system theory.

This chapter has emphasized that the new challenges will require some adaptations of the current AGC strategies to satisfy the general needs of the different market organizations and the specific characteristics of each power system. The existing market-based AGC configurations and new concepts were briefly discussed, and an updated frequency response model for decentralized AGC markets was introduced.

References

1. H. Bevrani. 2009. *Robust power system frequency control*. New York: Springer.
2. H. Bevrani. 2004. Decentralized robust load-frequency control synthesis in restructured power systems. PhD dissertation, Osaka University.
3. E. Hirst, B. Kirby. 1999. Separating and measuring the regulation and load-following ancillary services. *Utilities Policy* 8(2):75–81.
4. Power Systems Engineering Research Center (PSERC). 2009. *Impact of increased DFIG wind penetration on power systems and markets*. Final project report. PSERC, Arizona State University, Phoeniz, AZ, USA.

5. J. Kumar, N. G. K. Hoe, G. B. Sheble. 1997. AGC simulator for price-based operation. Part I. A model. *IEEE Trans. Power Syst.* 2(12):527–32.
6. J. Kumar, N. G. K. Hoe, G. B. Sheble. 1997. AGC simulator for price-based operation. Part II. Case study results. *IEEE Trans. Power Syst.* 2(12):533–38.
7. M. Shahidehpour, H. Yamin, Z. Li. 2002. *Market operations in electric power systems: Forecasting, scheduling, and risk management.* New York: John Wiley & Sons.
8. Y. Rebours. 2008. A comprehensive assessment of markets for frequency and voltage control ancillary services. PhD dissertation, University of Manchester.
9. W. Hogan. 2002. Electricity market design and structure: Working paper on standardized transmission services and wholesale market design. Washington, DC. http://www.ferc.fed.us.
10. R. D. Chritie, A. Bose. 1996. Load frequency control issues in power system operation after deregulation. *IEEE Trans. Power Syst.* 11(3):1191–200.
11. S. Bhowmik, K. Tomsovic, A. Bose. 2004. Communication models for third party load frequency control. *IEEE Trans. Power Syst.* 19(1):543–48.
12. H. Bindner, O. Gehrke. 2009. *System control and communication.* Risø Energy Report 8, 39–42. http://130.226.56.153/rispubl/reports/ris-r-1695.pdf.
13. Y. Rebours, D. Kirschen. 2005. *A survey of definitions and specifications of reserve services.* Technical report, University of Manchester. http://www.eee.manchester.ac.uk/research/groups/eeps/publications/reportstheses/aoe/rebours%20et%20al_tech%20rep_2005B.pdf.
14. G. A. Chown, B. Wigdorowitz. 2004. A methodology for the redesign of frequency control for AC networks. *IEEE Trans. Power Syst.* 19(3):1546–54.
15. UCTE. 2009. *UCTE operation handbook.* http://www.ucte.org.
16. B. Delfino, F. Fornari, S. Massucco. 2002. Load-frequency control and inadvertent interchange evaluation in restructured power systems. *IEE Proc. Gener. Transm. Distrib.* 149(5):607–14.
17. G. Dellolio, M. Sforna, C. Bruno, M. Pozzi. 2005. A pluralistic LFC scheme for online resolution of power congestions between market zones. *IEEE Trans. Power Syst.* 20(4):2070–77.
18. I. Egido, F. Fernandez-Bernal, L. Rouco. 2009. The Spanish AGC system: Description and analysis. *IEEE Trans. Power Syst.* 24(1):271–78.
19. L. Olmos, J. I. Fuente, J. L. Z. Macho, R. R. Pecharroman, A. M. Calmarza, J. Moreno. 2004. New design for the Spanish AGC scheme using an adaptive gain controller. *IEEE Trans. Power Syst.* 19(3):1528–37.
20. N. Maruejouls, T. Margotin, M. Trotignon, P. L. Dupuis, J. M. Tesseron. 2000. Measurement of the load frequency control system service: Comparison between American and European indicators. *IEEE Trans. Power Syst.* 15(4):1382–87.
21. I. Arnott, G. Chown, K. Lindstrom, M. Power, A. Bose, O. Gjerde, R. Morfill, N. Singh. 2003. Frequency control practices in market environments. In *Quality and Security of Electric Power Delivery Systems 2003, CIGRE/IEEE PES International Symposium,* Montreal, QC, ON, 143–48.
22. PSERC. 2008. Agent modelling for integrated power systems. Project report. http://www.pserc.org.
23. K. Bhattacharya, M. H. J. Bollen, J. E. Daalder. 2001. *Operation of restructured power systems.* Boston: Kluwer Academic Publishers.
24. H. Singh. 1999. Auctions for ancillary services. *Decision Support Systems* 24(3–4):183–91.

25. H.-P. Chao, R. Wilson. 2002. Multi-dimensional procurement auctions for power reserves: Robust incentive-compatible scoring and settlement rules. *J. Regulatory Econ.* 22(2):161–83.
26. G. Chicco, G. Gross. 2004. Competitive acquisition of prioritizable capacity-based ancillary services. *IEEE Trans. Power Syst.* 19(1):569–76.
27. F. D. Galiana, F. Bouffard, J. M. Arroyo, J. F. Restrepo. 2005. Scheduling and pricing of coupled energy and primary, secondary, and tertiary reserves. *Proc. IEEE* 93(11):1970–83.
28. J. J. Ford, H. Bevrani, G. Ledwich. 2009. Adaptive load shedding and regional protection. *Int. J. Elect. Power Energy Syst.* 31:611–18.
29. H. Bevrani, G. Ledwich, J. J. Ford, Z. Y. Dong. 2008. On power system frequency control in emergency conditions. *J. Elect. Eng. Technol.* 3(4):499–508.
30. Frontier Economics and Consentec. 2005. *Benefits and practical steps towards the integration of intraday electricity markets and balancing mechanisms.* London: Frontier Economics Ltd. http://europa.eu.int/comm/energy/electricity/publications/doc/frontier_consentec_balancing_dec_2005.pdf.
31. V. Donde, M. A. Pai, I. A. Hiskens. 2001. Simulation and optimization in a AGC system after deregulation. *IEEE Trans. Power Syst.* 16(3):481–89.
32. H. Bevrani, Y. Mitani, K. Tsuji. 2004. Robust AGC: Traditional structure versus restructured scheme. *IEEJ Trans. Power Energy* 124-B(5):751–61.
33. H. Bevrani, Y. Mitani, K. Tsuji, H. Bevrani. 2005. Bilateral-based robust load-frequency control. *Energy Conversion Management* 46:1129–46.
34. PSERC. 2005. New system control methodologies: Adapting AGC and other generator controls to the restructured environment. Project report. http://www.pserc.org.
35. J. M. Arroyo, A. J. Conejo. 2002. Optimal response of a power generator to energy, AGC, and reserve pool-based markets. *IEEE Trans. Power Syst.* 17(2):404–10.
36. H. Bevrani, Y. Mitani, K. Tsuji. 2004. Robust decentralized AGC in a restructured power system. *Energy Conversion Management* 45:2297–312.

5

Neural-Network-Based AGC Design

Recent achievements on artificial neural networks (ANNs) promote great interest principally due to their capability to learn to approximate well any arbitrary nonlinear functions, and their ability for use in parallel processing and multivariable systems. The capabilities of such networks could be properly used in the design of adaptive control systems. In this chapter, following an introduction of ANN application in control systems, an approach based on artificial flexible neural networks (FNNs) is proposed for the design of an AGC system for multiarea power systems in deregulated environments.

Here, the power system is considered a collection of separate control areas under the bilateral scheme. Each control area that is introduced by one or more distribution companies can buy electric power from some generation company to supply the area load. The control area is responsible for performing its own AGC by buying enough regulation power from prespecified generation companies, via an FNN-based supplementary control system.

The proposed control strategy is applied to a single- and three-control area power systems. The resulting controllers are shown to minimize the effect of disturbances and achieve acceptable frequency regulation in the presence of various load change scenarios.

5.1 An Overview

In a deregulated environment, an AGC system acquires a fundamental role to enable power exchanges and to provide better conditions for the electricity trading. The AGC is treated as an ancillary service essential for maintaining the electrical system reliability at an adequate level. Technically, this issue will be more important as independent power producers (IPPs), renewable energy source (RES) units, and microgrid networks get into the electric power markets.[1]

As mentioned in Chapter 4, there are several schemes and organizations for the provision of AGC services in countries with a restructured electric industry, differentiated by how free the market is, who controls generator units, and who has the obligation to execute AGC. Some possible AGC structures are introduced in Chapter 4. Under a deregulated environment, several notable solutions have already been proposed.[1,2] Here, it is assumed that in each control area, the necessary hardware and communication facilities to enable reception of data and control signals are available, and Gencos can bid up and down regulations by price and MW-volume for each predetermined time period to the regulating market. Also, the control center can distribute load demand signals to available generating units on a real-time basis.

The participation factors, which are actually time-dependent variables, must be computed dynamically based on the received bid prices, availability, congestion problem, and other related costs, in the case of using each applicant (Genco). Each participating unit will receive its share of the demand, according to its participation factor, through a dynamic controller that usually includes a simple proportional-integral (PI) structure in a real-world power system. Since the PI controller parameters are usually tuned based on classical experiences and trial-and-error approaches, they are incapable of obtaining good dynamical performance for a wide range of operating conditions and various load scenarios.

An appropriate computation method for the participation factors and desired optimization algorithms for the AGC systems has already been reported in Bevrani.[1] Several intelligent-based AGC schemes are also explained in Chapter 3. In continuation, this chapter focuses on the design of a dynamic controller unit using artificial FNNs. Technically, this controller, which is known as a supplementary control unit, has an important role to guarantee a desired AGC performance. An optimal design ensures smooth coordination between generator set point signals and the scheduled operating points. This chapter shows that the FNN control design provides an effective design methodology for the supplementary frequency controller synthesis in a new environment.

It is notable that this chapter is not about how to price either energy or any other economical aspects and services. These subjects are briefly addressed in Chapter 4. It is assumed that the necessary pricing mechanism and congestion management program are established by either free markets, a specific government regulation, or voluntary agreements, and this chapter only focuses on a technical solution for designing supplementary control loops in a bilateral-based electric power market.

The ANNs have already been used to design an AGC system for a power system with a classical (regulated) structure.[3–12] Generally, in all applications, the learning algorithms cause the adjustment of the connection weights so that the controlled system gives a desired response. The most common ANN-based AGC structures are briefly explained in Chapter 3. In this chapter, in order to achieve a better performance, the FNN-based AGC system has been proposed with dynamic neurons that have wide ranges of variation.[13]

The proposed control strategy is applied to a three-control area example. The obtained results show that designed controllers guarantee the desired performance for a wide range of operating conditions. This chapter is organized as follows. An introduction on ANN-based control systems with commonly used configurations is given in Section 5.2. The ANN with flexible neurons to perform FNN is described in Section 5.3. Section 5.4 presents the bilateral AGC scheme and dynamical modeling. The FNN-based AGC framework is given in Section 5.5. In Section 5.6, the proposed strategy is applied to a single- and three-control area examples, and some simulation results are presented.

5.2 ANN-Based Control Systems

For many years, it was a dream of scientists and engineers to develop intelligent machines with a large number of simple elements, such as neurons in biological organisms. McCulloch and Pitts[14] published the first systematic study of the artificial neural network. In the 1950s and 1960s, a group of researchers combined these biological and psychological insights to produce the first ANN.[15,16] Further investigations in ANN continued before and during the 1970s by several pioneer researchers, such as Rosenblatt, Grossberg, Kohonen, Widrow, and others. The primary factors for the recent resurgence of interest in the area of neural networks are dealing with learning in a complex, multilayer network, and a mathematical foundation for understanding the dynamics of an important class of networks.[17]

The interest in ANNs comes from the networks' ability to mimic the human brain as well as its ability to learn and respond. As a result, neural networks have been used in a large number of applications and have proven to be effective in performing complex functions in a variety of fields, including control systems. Adaptation or learning is a major focus of neural net research that provides a degree of robustness to the ANN model.

5.2.1 Fundamental Element of ANNs

As mentioned in Chapter 3, an ANN consists of a number of nonlinear computational processing elements (neurons), arranged in several layers, including an input layer, an output layer, and one or more hidden layers in between. Every layer usually contains several neurons, and the output of each neuron is usually fed into all or most of the inputs of the neurons in the next layer. The input layer receives input signals, which are then transformed and propagated simultaneously through the network, layer by layer.

The ANNs are modeled based on biological structures for information processing, including specifically the nervous system and its basic unit, the neuron. Signals are propagated in the form of potential differences between the inside and outside of cells. Each neuron is composed of a body, one axon, and a multitude of dendrites. Dendrites bring signals from other neurons into the cell body (soma). The cell body of a neuron sums the incoming signals from dendrites as well as the signals from numerous synapses on its surface. Once the combined signal exceeds a certain cell threshold, a signal is transmitted through the axon. However, if the inputs do not reach the required threshold, the input will quickly decay and will not generate any action. The axon of a single neuron forms synaptic connections with many other neurons. Cell nonlinearities make the composite action potential a nonlinear function of the combination of arriving signals.

A mathematical model of the neuron is depicted in Figure 5.1, which shows the basic element of an ANN. It consists of three basic components that include weights W_j, threshold (or bias) θ, and a single activation function $f(\cdot)$. The values $W_1, W_2, ..., W_n$ are weight factors associated with each node to determine the strength of input row vector $X^T = [x_1\, x_2\, ...\, x_n]$. Each input is multiplied by the associated weight of the neuron connection. Depending upon the activation function, if the weight is positive, the resulting signal commonly excites the node output, whereas for negative weights it tends to inhibit the node output.

The node's internal threshold θ is the magnitude offset that affects the activation of the node output y as follows:

$$y(k) = f\left(\sum_{j=1}^{n} W_j x_j(k) + W_0 \theta\right) \qquad (5.1)$$

This network, which is a simple computing element, was called the perceptron by Rosenblatt in 1959, which is well discussed in Haykin.[18] The nonlinear cell function (activation function) can be selected according to the application. Sigmoid functions are a general class of monotonically nondecreasing functions taking on bounded values. It is noted that as the threshold or bias changes, the activation functions may also shift. For many ANN training algorithms, including backpropagation, the derivative of $f(\cdot)$ is needed so that the activation function selected must be differentiable.

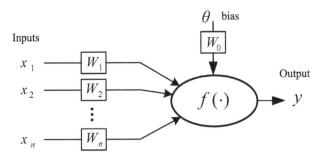

FIGURE 5.1
Basic element of an artificial neuron.

5.2.2 Learning and Adaptation

New neural morphologies with learning and adaptive capabilities have infused new control power into the control of complex dynamic systems. *Learning* and *adaptation* are the two keywords associated with the notion of ANNs. Learning and adaptation in the field of intelligent control systems were basically introduced in the early 1960s, and several extensions and advances have been made since then. In 1990, Narendra and Parthasarathy suggested that feedforward ANNs could also be used as components in feedback systems. After that, there were a lot of activities in the area of ANN-based identification and control, and a profusion of methods was suggested for controlling nonlinear and complex systems. Although much of the research was heuristic, it provided empirical evidence that ANNs could outperform traditional methods.[19]

There exist different learning algorithms for ANNs, but they normally encounter some technical problems, such as local minimization, low learning speed, and high sensitivity to initial conditions, among others. Recently, some learning algorithms based on other powerful tools, such as Kalman filtering, have been proposed.[20]

Recent advances in static and dynamic ANN have created a profound impact on the neural architecture for adaptation and control, introduction to the backpropagation algorithms, and identification and control problems for a general class of dynamic systems. The subject of adaptive control systems, with various terms such as *neoadaptive, intelligent,* and *cognitive control systems,* falls within the domain of control of complex industrial systems with reasoning, learning, and adaptive abilities.[21]

The backpropagation learning algorithm is known as one of the most efficient learning procedures for multilayer ANNs. One reason for wide use of this algorithm, which will be described later, is its simplicity. These learning algorithms provide a special attribute for the design and operation of the dynamic ANN for a given task, such as design of controllers for complex dynamic systems. There are several approaches for deriving the backpropagation algorithm. The simplest derivation is presented in Fogelman-Soulie et al.[22] and LeCun.[23] This approach is directly influenced from the optimal control theory, which uses Lagrange multipliers to obtain the optimal values of a set of control variables.

Direct analytic computation and recursive update techniques are two basic learning approaches to determining ANN weights. Learning algorithms may be carried out in continuous time or discrete time via differential or difference equations for the weights. There are many learning algorithms, which can be classified into three categories: (1) a supervised learning algorithm uses a supervisor that knows the desired outcomes and tunes the weights accordingly; (2) an unsupervised learning algorithm uses local data, instead of a supervisor, according to emergent collective properties; and (3) a reinforcement learning algorithm uses some reinforcement signal, instead of the output error of that neuron, to tune the weights.

Unlike the unsupervised learning, both supervised and reinforcement learning algorithms require a supervisor to provide training signals in different ways. In supervised learning, an explicit signal is provided by the supervisor throughout to guide the learning process, while in reinforcement learning, the role of the supervisor is more evaluative than instructional.[24] An algorithm for reinforcement learning with its application in multiagent-based AGC design is described in Chapter 7.

The principles of reinforcement learning and related ideas in various applications have been extended over the years. For example, the idea of *adaptive critic*[25] was introduced as an extension of the mentioned general idea of reinforcement learning in feedback control systems. The adaptive critic ANN architecture uses a critic ANN in a high-level supervisory capacity that critiques the system performance over time and tunes a second action ANN (for generating the critic signal) in the feedback control loop. The critic ANN can select either the standard Bellman equation, Hamilton-Jacobi-Bellman equation, or a simple weighted sum of tracking errors as the performance index, and it tries to minimize the index. In general, the critic conveys much less information than the desired output required in supervisory learning.[26] Some recent research works use a supervisor in the actor-critic architecture, which provides an additional source of evaluation feedback.[27]

On the other hand, the learning process in an ANN could be offline or online. Offline learning is useful in feedforward applications such as classification and pattern recognition, while in feedback control applications, usually online learning, which is more complex, is needed. In an online learning process, the ANN must maintain the stability of a dynamical system while simultaneously learning and ensuring that its own internal states and weights remain bounded.

5.2.3 ANNs in Control Systems

Serving as a general way to approximate various nonlinear static and dynamic relations, ANN has the ability to be easily implemented for complex control systems. While, in most cases, only simulations supported the proposed control ideas, presently more and more theoretical results are proving the soundness of neural approximation in control systems.[28]

Applications of ANNs in feedback control systems are basically distinct from those in open-loop applications in the fields of classification, pattern recognition, and approximation of nondynamic functions. In latter applications, ANN usage has developed over the years to show how to choose network topologies and select weights to yield guaranteed performance. The issues associated with weight learning algorithms are well understood. In ANN-based feedback control of dynamical systems, the problem is more complicated, and the ANN must provide stabilizing controls for the system as well as ensure that all its weights remain bounded.

The main objective of intelligent control is to implement an autonomous system that can operate with increasing independence from human actions in an uncertain environment. This objective could be achieved by learning from the environment through a feedback mechanism. The ANN has the capability to implement this kind of learning. Indeed, an ANN consists of a finite number of interconnected neurons (as described earlier) and acts as a massively parallel distributed processor, inspired from biological neural networks, which can store experimental knowledge and make it available for use.

The research on neural networks promotes great interest principally due to the capability of static ANNs to approximate arbitrarily well any continuous function. The most used ANN structures are *feedforward* and *recurrent* networks. The latter offers a better-suited tool for intelligent feedback control design systems.

The most proposed ANN-based control systems use six general control structures that are conceptually shown in Figure 5.2: (1) using the ANN system as a controller to provide a direct control command signal in the main feedback loop; (2) using ANN for tuning the parameters of the existing fixed structure controller (I, PI, PID, etc.); (3) using the ANN system as an additional controller in parallel with the existing conventional controller, such as I, PI, and PID, to improve the closed-loop performance; (4) using the ANN controller with an additional intelligent mechanism to control a dynamical plant; (5) using ANN in a feedback control scheme with an intelligent recurrent observer/identifier; and finally, (6) using ANNs in an adaptive critic control scheme. The above six configurations are presented in Figure 5.2a–f, respectively.

In all control schemes, the ANN collects information about the system response, adjusts weights via a learning algorithm, and recommends an appropriate control signal. In Figure 5.2b, the ANN performs an automatic tuner. The main components of the ANN as an intelligent tuner for other (conventional) controllers (Figure 5.2b) include a response recognition unit to monitor the controlled response and extract knowledge about the performance of the current controller gain setting, and an embedded unit to suggest suitable changes to be made to the controller gains. One can use a linear model predictive controller (MPC) instead of a conventional controller.[29] In this case, the combination of a linear MPC and ANN unit in Figure 5.2b represents a nonlinear MPC. The ANN unit provides an estimate for the deviation between the predicted value of the output computed via the linear model and the actual nonlinear system output, at a giving sampling time.

The structure shown in Figure 5.2e is mainly useful for trajectory tracking control problems. A recurrent high-order ANN observer can be used to implement the intelligent observer/identifier block.[30] An adaptive critic control scheme, shown in Figure 5.2f, is comprised of a critic ANN and an action ANN that approximate the global control based on the nonlinear plant and its model. The critic ANN evaluates the action ANN performance by analyzing predicted states (from the plant model) and real measurements (from the actual plant). The adaptive critic control designs have the

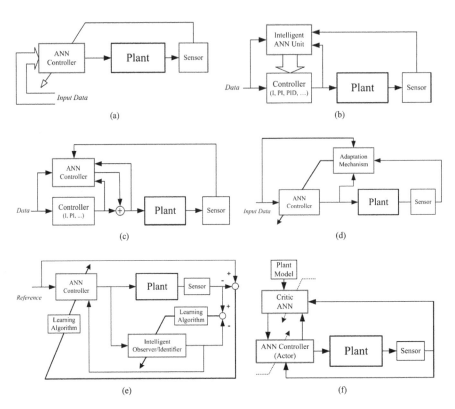

FIGURE 5.2
Popular configurations for ANN-based control systems in (a)–(f).

potential to replace critical aspects of brain-like intelligence, that is, the ability to cope with a large number of variables in a real environment for ANNs. The origins of the adaptive critical control are based on the ideas of synthesized from reinforcement learning, real-time derivation, backpropagation, and dynamic programming. An application study for the mentioned control structure, entitled the dual heuristic programming adaptive critic control method, is presented in Ferrari.[31]

In addition to the above control configurations, ANN is widely used as a plant identifier in control systems. There are two structures for plant identification: the forward and inverse structures. In case of the forward configuration, the ANN receives the same input as the plant, and the difference between the plant output and the ANN output is minimized usually using the backpropagation algorithm. But, the inverse plant identification employs the plant output as the ANN input, while the ANN generates an approximation of the input vector of the real plant.

When an ANN is used as a controller, most of the issues are similar to those of the identification case. The main difference is that the desired output of

ANN controller, that is, the appropriate control input to be fed to the plant, is not available but has to be induced from the known desired plant output. In order to achieve this, one uses either approximations based on a mathematical model of the plant or an ANN identifier.[29] To perform an adaptive control structure, usually ANNs are combined to both identification and control parts, as shown in Figure 5.2e. *Internal model control* can be considered as an application for this issue. Such a design, which is schematically shown in Figure 5.3, is robust against model inaccuracies and plant disturbances.[28] An idea of the internal model control[32] consists of employing a model of the plant and modifying the reference signal $r^*(k) = r(k) - (y(k) - \tilde{y}(k))$, where $\tilde{y}$ represents the internal model output.

Over the past years, ANNs have been effectively used for regulation and tracking problems, which are two control problems of general interest. Regulation involves the generalizations of a control input that stabilize the system around an equilibrium state. In the tracking problem, a reference output is specified and the output of the plant is to approximate it in some sense with as small a difference error as possible. For theoretical analysis, this error is assumed to be zero, so that asymptotic tracking is achieved. Numerous problems are encountered when an ANN is used to control a dynamical system, including regulation and tracking issues. Some problems can be briefly considered as follows:

1. Since the ANN is in a feedback loop with the controlled plant, dynamic rather than static backpropagation is needed to adjust the parameters along the negative gradient. However, in practice, only static backpropagation is used.[19]

2. In most control applications, an approximate model of the plant is needed, and to improve the performance further, the parameters of an additional neural network may be adjusted, as in point *i*.

3. Because of the complexity of the structure of a multilayer ANN and the nonlinear dependence of its map on its parameter values, stabil-

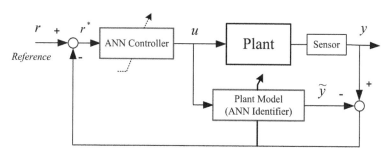

FIGURE 5.3
Internal model control scheme.

ity analysis of the resulting system is always very difficult and quite often intractable.

4. In practice, the three-step procedure described earlier is used in industrial problems, to avoid adaptation online and the ensuing stability problems. However, if the feedback system is stable, online adaptive adjustments can improve performance significantly, provided such adjustments are small.[19]

5. As an important part of ANN-based control system design, the quality of selection methods for ANN initial conditions significantly influences the quality of the control solution.

Selection of initial conditions in an ANN-based control system is also known as an important issue. In multiextremum optimization problems, some initial values may not guarantee the achievement of extremum with a satisfactory value of optimization functional. The initial conditions are usually selected according to the a priori information about distributions at the already known structure of the open-loop ANN and selected optimization structure (control strategy). Methods for the selection of initial conditions can be classified into three categories according to the form of the used information:[33] random initial conditions without use of the learning sample, deterministic initial conditions without use of the learning sample, and initial conditions with use of the learning sample.

5.3 Flexible Neural Network

5.3.1 Flexible Neurons

As mentioned in the previous section, the activation function is the most important part of a neuron (Figure 5.1), and it is usually modeled using a sigmoid function. The flexibility of ANNs can be increased using flexible sigmoid functions (FSFs). Basic concepts and definitions of the introduced FSF were described in Teshnehlab and Watanabe.[13] The following hyperbolic tangent function as a sigmoid unit function is considered in hidden and output layers:

$$f(x,a) = \frac{1-e^{-2xa}}{a(1+e^{-2xa})} \tag{5.2}$$

The shape of this bipolar sigmoid function can be altered by changing the parameter a, as shown in Figure 5.4. It also has the property

$$\lim_{a \to 0} f(x,a) = x \tag{5.3}$$

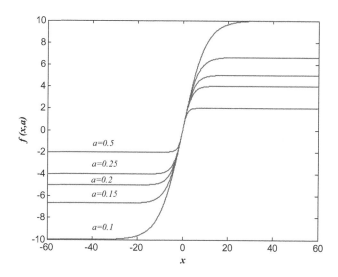

FIGURE 5.4
Sigmoid function with changeable shape.

Thus, it is proved that the previous function becomes linear when $a \to 0$, while the function becomes nonlinear for large values of a.[13] It should be noted that in this study, the learning parameters are included in the update of connection weights and sigmoid function parameters (SFPs). Generally, the main idea is to present an input pattern, allow the network to compute the output, and compare this to the desired signals provided by the supervisor or reference signal. Then, the error is utilized to modify connection weights and SFPs in the network to improve its performance with minimizing the error, as a flexible neural network (FNN) system.

5.3.2 Learning Algorithms in an FNN

The learning process of FNNs for control area i is to minimize the performance function given by

$$J = \frac{1}{2}(y_{di} - y_i^M)^2 \tag{5.4}$$

where y_{di} represents the reference signal, y_i^M represents the output unit, and M denotes the output layer. It is desirable to find a set of parameters in the connection weights and SFPs that minimize the J, considering the same input-output relation between the layer k and the layer $(k + 1)$. It is useful to consider how the error varies as a function of any given connection weights and SFPs in the system.

The error function procedure finds the values of all of the connection weights and SFPs that minimize the error function using a gradient descent method. That is, after each pattern has been presented, the error gradient moves toward its minimum for that pattern, provided a suitable learning rate. Learning of SFPs by employing the gradient descent method, the increment of a_i^k denoted by Δa_i^k, can be obtained as

$$\Delta a_i^k = -\eta_1 \frac{\partial J}{\partial a_i^k} \qquad (5.5)$$

where $\eta_1 > 0$ is a *learning rate* given by a small positive constant. In the output layer M, the partial derivative of J with respect to a is described as follows:[38]

$$\frac{\partial J}{\partial a_i^M} = \frac{\partial J}{\partial y_i^M} \frac{\partial y_i^M}{\partial a_i^M} \qquad (5.6)$$

Here, defining

$$\sigma_i^M \equiv -\frac{\partial J}{\partial y_i^M} \qquad (5.7)$$

gives

$$\sigma_i^M = (y_{di} - y_i^M) \qquad (5.8)$$

The next step is to calculate a in the hidden layer k:

$$\frac{\partial J}{\partial a_i^k} = \frac{\partial J}{\partial y_i^k} \frac{\partial y_i^k}{\partial a_i^k} = \frac{\partial J}{\partial y_i^k} f^*(h_i^k, a_i^k) \qquad (5.9)$$

where h denotes outputs of the hidden layer, and by defining

$$a_i^k \equiv -\frac{\partial J}{\partial y_i^k} \qquad (5.10)$$

we have

$$\frac{\partial J}{\partial y_i^k} = \sum_m \frac{\partial J}{\partial y_m^{k+1}} \frac{\partial y_m^{k+1}}{\partial y_i^k}$$

$$= -\sum_m \sigma_m^{k+1} \frac{\partial y_m^{k+1}}{\partial h_m^{k+1}} \frac{\partial h_m^{k+1}}{\partial y_i^k} = -\sum_m \sigma_m^{k+1} \frac{\partial f(h_m^{k+1}, a_m^{k+1})}{\partial h_m^{k+1}} w_{i,m}^{k,k+1} \qquad (5.11)$$

where

$$a_i^k = \sum_m \sigma_m^{k+1} \frac{\partial f(h_m^{k+1}, a_m^{k+1})}{\partial h_m^{k+1}} w_{i,m}^{k,k+1} \qquad (5.12)$$

Therefore, the learning update equation for a in the output and hidden layer neurons is obtained, respectively, as follows:

$$a_i^k(t+1) = a_i^k(t) + \eta_1 \sigma_i^k f^*(h_i^k, \alpha_i^k) + \alpha_1 \Delta a_i^k(t) \qquad (5.13)$$

where $f'(.,.)$ is defined by $\partial f(.,a_i{}^M)/\partial y_i{}^M$ in the output layer and $\partial f(.,a_i{}^k)/\partial y_i{}^k$ in the hidden layer, and α_1 is the *stabilizing coefficient* defined by $0 < \alpha_1 < 1$. For deeper insights into the subject, interested readers are referred to Teshnehlab and Watanabe.[13]

Generally, the learning algorithm of connection weights has been studied by different authors. Here, it can be simply summarized as follows:

$$w_{ij}{}^{k,\,k-1}(t+1) = w_{ij}{}^{k,\,k-1}(t) + \eta_2 \delta_j^k y_j{}^{k-1} + \alpha_2 \Delta w_{ij}{}^{k,\,k-1}(t) \tag{5.14}$$

where t denotes the tth update time, $\eta_2 > 0$ is a learning rate given by a small positive constant, α_2 is a stabilizing (or momentum) coefficient defined by $0 < \alpha_2 < 1$, and

$$\delta_j^M = (y_{dj} - y_j{}^M) f'(h_j{}^M) \tag{5.15}$$

$$\delta_j^k = f'(h_j{}^k) \sum_m \delta_m^{k+1} w_{ij}{}^{k,\,k+1} \tag{5.16}$$

$$f'(h_j{}^M) = df'(h_j{}^M)/dh_j{}^M \tag{5.17}$$

5.4 Bilateral AGC Scheme and Modeling

5.4.1 Bilateral AGC Scheme

As mentioned in Chapter 4, depending on the electrical system structure, there are different control frameworks and AGC/LFC schemes, but a common objective is restoring the frequency and the net interchanges to their nominal values, in each control area. The present work studies AGC design in a given control area under bilateral structure, similar to the decentralized pluralistic scheme defined by UCTE[34] and described in Chapter 4.

In a bilateral AGC scheme, the power system is considered a collection of distribution control areas interconnected through high-voltage transmission lines or tie-lines. Each control area has its own load frequency controller and is responsible for tracking its own load and honoring tie-line power exchange contracts with its neighbors. Similar to Chritie and Bose,[35] the general theme in our chapter is that the loads (Discos) are responsible for purchasing the ancillary services they require. Contractual agreements, between regulation power producers (Gencos) and consumers (Discos), are signed together with transmission access license energy transactions. In the assumed bilateral AGC structure, each distribution area purchases regulating power from one

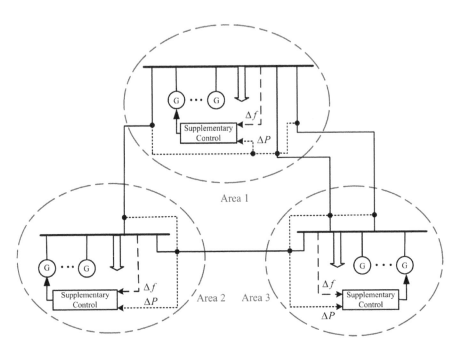

FIGURE 5.5
Three control areas.

or more Gencos. A separate control process exists for each control area. Such
a configuration for a three-control area power system is conceptually shown
in Figure 5.5.

The control areas regulate their frequency by their own supplementary control loop. According to the UCTE definitions, a control area is the smallest portion of a power system equipped with an autonomous AGC system. A control
block may be formed by one or more control areas working together to meet a
specified AGC requirement with respect to the neighboring control blocks.[37]

If some control areas perform a control block, which is not applied in the
present example, in this case a block coordinator (market operator) coordinates
the whole block toward its neighbor blocks/control area by means of its own
supplementary controller and regulating capacity.[34,36] The control algorithm for
each control area is executed at the generating units, since ultimately the Gencos
must adjust the governor set points of generators for the AGC purpose.

5.4.2 Dynamical Modeling

In most ANN applications, specifically for performing the backpropagation
linear algorithm, an approximate dynamic model of the plant is needed. This
model can be obtained by either identification methods or mathematical differential equations. For this purpose, differential equations of an aggregated
control area model are used.

Consider that a general distribution control area includes N generator companies that supply the area load, and assume the kth Genco G_k is able to generate enough power to satisfy the necessary participation factor for tracking the load and performing the AGC task, and other Gencos are the main supplier for area load.

Although power systems are inherently nonlinear, for intelligent AGC design, a simplified (and linearized) model is usually used. In intelligent control strategies (such as the one considered in this chapter), the error caused by the simplification and linearization can be considered in respect to their self-tuning and adaptive property. For simplicity, assume that each Genco has one generator. The linearized dynamics of the individual generators are given by

$$\frac{2H_i}{f_0}\frac{d\Delta f_i}{dt} = \Delta P_{mi} - \Delta P_{Li} - \Delta P_{di} - D_i\Delta f_i$$

$$\frac{d\Delta\delta_i}{dt} = 2\pi\Delta f_i \qquad\qquad , i = 1, 2, ..., N \qquad (5.18)$$

where H_i is the constant of inertia, D_i is the damping factor, δ_i is the rotor angle, f_0 is the nominal frequency, P_m is the turbine (mechanical) power, P_{di} is the disturbance (power quantity), and P_{Li} is the electrical power.

The generators are equipped with a speed governor. Assuming nonreheat steam type for the generating units, the linear models of speed governors and turbines associated with generators are given by deferential equations:[1]

$$\frac{d\Delta P_{gi}}{dt} = -\frac{1}{T_{gi}}\Delta P_{gi} + \frac{K_{gi}}{T_{gi}}(\Delta P_{Ci} - \frac{1}{R_i}\Delta f_i)$$

$$\frac{d\Delta P_{mi}}{dt} = -\frac{1}{T_{ti}}\Delta P_{mi} + \frac{K_{mi}}{T_{ti}}\Delta P_{gi} \qquad , i = 1, ..., N \qquad (5.19)$$

where P_g is the steam valve power, R_i is the droop characteristic, T_t and T_g are the time constants of the turbine and governor, K_m and K_t are the gains of the turbine and governor, and finally, P_{Ci} is the reference set point (control input).

The individual generator models are coupled to each other via the system network. Mathematically, the local state space of each individual generator must be extended to include the system coupling variable (such as rotor angle), which allows the dynamics at one point on the system to be transmitted to all other points. The state space model of a control area can be easily obtained as follows:[38]

$$\dot{x}_i = A_i x_i + B_i u_i + F_i w_i$$

$$y_i = C_i x_i + E_i w_i \qquad\qquad (5.20)$$

5.5 FNN-Based AGC System

Here, the objective is to formulate the AGC problem in each control area and propose an effective supplementary control loop based on the ANNs. Since there is a strong relationship between the training of ANNs and adaptive/self-tuning control, increasing the flexibility of structure induces a more efficient learning ability in the system, which in turn causes less iteration and better error minimization. To obtain the improved flexibility, teaching signals and other parameters of ANNs (such as connection weights) should be related to each other.

Here, a sigmoid unit function, as a mimic of the prototype unit, to give a flexible structure to the neural network, is used. For this purpose, a hyperbolic tangential form of the sigmoid unit function, with a parameter that must be learned,[13] is introduced to fulfill the above-mentioned goal. The overall scheme of the proposed control system for a given control area is shown in Figure 5.6.

The FNN uses a backpropagation algorithm in a supervised learning mode. The accuracy and speed of the backpropagation method is improved using dynamic neurons. The main idea is to modify connection weights and SFPs in the proposed FNN-based controller to minimize the output error (e) signal and improve system performance. On the other hand, it is desirable to find a set of parameters in the connection weights and SFPs that minimize the output error signal.

The designed controller acts to maintain area frequency and total exchange power close to the scheduled value by sending a corrective signal to the assigned Gencos ($\Delta P_{Ci} = u_i$). This supplementary regulating power signal, weighted by the generator participation factor α_{ij}, is used to modify the set points of generators. The I_i is an input vector that includes a property set

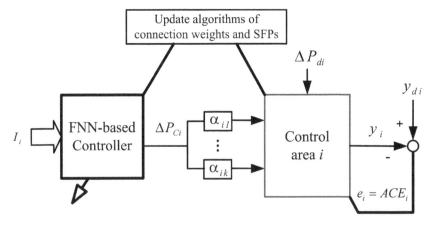

FIGURE 5.6
The overall synthesis framework.

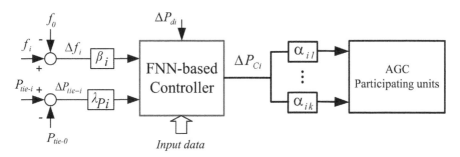

FIGURE 5.7
The proposed AGC scheme.

of output (y_i), reference (y_{di}), and control signal (u_i) in current and previous iteration steps.

The proposed control framework, shown in Figure 5.7, uses the tie-line power change (ΔP_{tie}) and frequency deviation (Δf) as main input signals. β_i and λ_{Pi} are properly set up coefficients of the supplementary control feedback system. ΔP_d represents the system disturbances. More measurement signals and data may need to feed the FNN control system as additional inputs (such as previous values of the control signal). As there are many Gencos in each area, the control signal has to be distributed among them in proportion to their participation in the AGC. Hence, the generator participation factor shows the sharing rate of each participant generator unit in the AGC task.

The general structure of an FNN-based controller is shown in Figure 5.8. In this figure, unit functions in the hidden and output layers are flexible functions. The number of hidden layers and the units in each layer are entirely dependent on the control area system, and there is no mathematical approach to obtain the optimum number of hidden layers,[17,39] since such selection generally falls into the application-oriented category.

The number of hidden layers can be chosen based on the training of the network using various configurations. It has been shown that the FNN configuration gives fewer hidden layers and nodes than traditional ANN,[8] which still yields the minimum root mean squares (RMS) error quickly and efficiently. Experiences and simulation results show that using a single hidden layer is sufficient to solve the AGC problem for many control areas with different structures.[38]

The equivalent discrete time-domain state space of the control area model (Equation 5.20) can be obtained as follows:

$$x_i(k+1) = A_i x_i(k) + B_i u_i(k) + F_i w_i(k)$$

$$y_i(k) = C_i x_i(k) + E_i w_i(k)$$

(5.21)

When the FNN controller is native, i.e., the network is with random initial weights and SFPs, an erroneous system input $u_i(k)$ may be produced by an

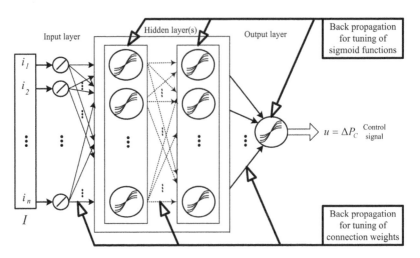

FIGURE 5.8
General structure of an FNN-based controller.

erroneous output $y_i(k)$. This output will then be compared with the reference signal $y_{di}(k)$. The resulting error signal $e_i(k)$ is used to train the weights and SFPs in the network using the backpropagation algorithm. In fact, the basic concept of the backpropagation method of learning is to combine a nonlinear perceptron-like system capable of making decisions with the objective error function and gradient descent method. With repetitive training, the network will learn how to respond correctly to the reference signal input.

As the amount of training increases, the network becomes more and more mature; hence, the area control error (ACE) will become smaller and smaller. However, as seen from Figure 5.8, the backpropagation of the error signal cannot be directly used to train the FNN controller. In order to properly adjust the weights and SFPs of the network using the backpropagation algorithm, the error (Equation 5.22) in the FNN controller output should be known.

$$e_i(k) = y_{di}(k) - y_i(k) = ACE_{di}(k) - ACE_i(k) = -ACE_i(k) \qquad (5.22)$$

where $u_{di}(k)$ is the desired driving input to the control area. Since only the system output error, as described in Equation 5.23, is measurable or available,

$$e_i(k) = y_{di}(k) - y_i(k) \qquad (5.23)$$

$\varepsilon_i(k)$ can be determined using the following expression:

$$\varepsilon_i(k) = e_i(k)\frac{\partial y_i(k)}{\partial u_i(k)} \qquad (5.24)$$

where the partial derivative in Equation 5.24 is the Jacobian of the control area model. Thus, the application of this scheme requires a through knowledge of the Jacobian of the system dynamical model. For simplicity, instead of Equation 5.22, one can use the following equation:

$$\varepsilon_i(k) = e_i(k) \frac{\Delta y_i(k) - \Delta y_i(k-1)}{\Delta u_i(k) - \Delta u_i(k-1)} \qquad (5.25)$$

This approximation avoids the introduction and training of a neural network emulator, which results in substantial savings in development time. The proposed supplementary feedback acts as a self-tuning controller that can learn from experience, in the sense that connection weights and SFPs are adjusted online; in other words, this controller should produce ever-decreasing tracking errors from sampling by using FNN.

5.6 Application Examples

A discrete time-domain state space model for control area i can be obtained as given in Equation 5.21. To achieve the AGC objectives, the proposed control strategy is applied to a control area as shown in Figure 5.9. As can be seen from the block diagram, a multilayer neural network including three layers is constructed. This network has nine units in the input layer, seven units in the hidden layer, and one unit in the output layer. The neural network acts as a controller to supply the AGC participation generating units with a correct driving input $u_i(k) = \Delta P_{Ci}(k)$, which is based on the reference input signal $y_{di}(k)$, previous system output signals $y_i(k-1)$, ..., $y_i(k-4)$, and control output signals $u_i(k-1)$, ..., $u_i(k-4)$. The $y_{di}(k)$ is the output variable $y_i(k)$, when the error signal must be equal to zero. Then the input vector of neural network is

$$I_i^T(k) = [y_{di}(k) \; y_i(k-1) \ldots y_i(k-4) \; u_i(k-1) \ldots u_i(k-4)] \qquad (5.26)$$

$h_{1i}, \ldots, h_{7i}$ are outputs of the hidden layer, and $u_i(k)$ is the output of the output layer of area i.

As shown in Figure 5.9, in the learning process not only the connection weights, but also the SFPs are adjusted. Adjusting the SFPs in turn causes a change in the shapes of sigmoid functions. The proposed learning algorithm considerably reduces the number of training steps, resulting in much faster training than with the traditional ANNs.[38]

For the problem at hand, simulations show that three layers are enough for the proposed FNN control system to obtain desired AGC performance. Increasing the number of layers does not significantly improve the control

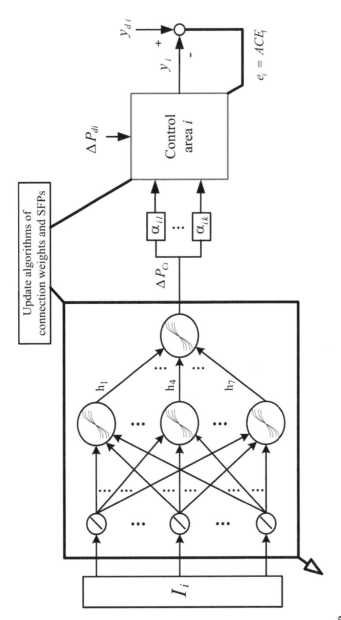

FIGURE 5.9
The proposed FNN controller for the control area *i*.

performance. In the proposed controller, the input layer uses the linear neurons, while the hidden and output layers use the bipolar FSFs given in Equation 5.2.

In order to demonstrate the effectiveness of the proposed strategy, some simulations were carried out. In these simulations, the proposed intelligent control design methodology is applied to a single- and three-control area power system example. The initial connection weights and initial uniform random number (URN) of sigmoid function unit parameters for each control area are properly chosen.

5.6.1 Single-Control Area

Consider a distribution control area and its suppliers, as shown in Figure 5.10, which perform a control area under the bilateral AGC policy. Gencos produce electric power that is delivered to the Disco either directly or through the Transco. In this example, the Disco buys firm power from Gencos 2, 3, and 4, and enough power from Genco 1 to supply its load and support the AGC system. It is assumed that Genco 1 is able to generate enough regulation power to satisfy the AGC needs. Transco delivers power from Genco 1, and it is also contracted to deliver power associated with the AGC problem. This control area is connected to its neighbors through L1 and L2 interconnection lines.

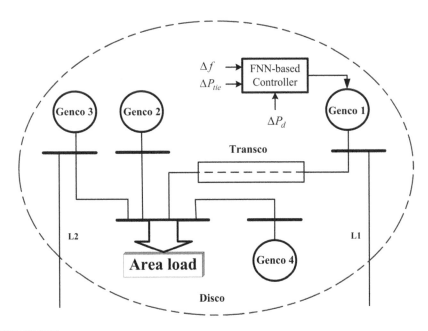

FIGURE 5.10
A single-control area.

TABLE 5.1

Applied Data for Simulation

Quantity	Genco 1	Genco 2	Genco 3	Genco 4
Rating (MW)	1,600	600	800	800
Constant of inertia: H (s)	5	4	4	5
Damping: D (pu MW/Hz)	0.02	0.01	0.01	0.015
Droop characteristic: $R(\%)$	4	5.2	5.2	5
Turbine's time constant: T_t	0.5	0.5	0.5	0.5
Governor's time constant: T_g	0.2	0.1	0.15	0.1
Gains: K_m, T_g	1	1	1	1

In the proposed structure, the Disco is responsible for tracking the load, and hence performing the AGC task by securing as much transmission and generation capacity as needed. Ultimately, the control algorithm is executed at Genco 1. It is assumed that each Genco has one generating unit. The power system parameters are given in Table 5.1. The sampling time is chosen as 1 ms. The initial values of sigmoid function parameters are randomly chosen from a domain of [0, 1], and the initial connection weights are considered as follows:

$$W1_0 = \begin{bmatrix} 0 & 1.085 & 1.084 & 1.083 & 1.082 & 0.031 & 0.031 & 0.032 & 0.032 \\ -0.0001 & -1.849 & -1.848 & -1.846 & -1.845 & -0.053 & -0.053 & -0.054 & -0.054 \\ 0.0001 & -3.404 & -3.401 & -3.398 & -3.395 & -0.097 & -0.098 & -0.099 & -0.099 \\ 0 & 3.448 & 3.445 & 3.442 & 3.44 & 0.098 & 0.099 & 0.1 & 0.099 \\ 0 & 1.484 & 1.483 & 1.481 & 1.48 & 0.042 & 0.043 & 0.043 & 0.043 \\ 0.0001 & 3.266 & 3.263 & 3.260 & 3.257 & 0.093 & 0.094 & 0.095 & 0.095 \\ -0.0001 & -4.073 & -4.07 & -4.066 & -4.063 & -0.116 & -0.117 & -0.118 & -0.118 \end{bmatrix}$$

$$W2_0 = \begin{bmatrix} -2.182 & 3.783 & 7.377 & -7.488 & -3.008 & -7.034 & 9.126 \end{bmatrix}$$

A system response for some simulation scenarios is shown in Figure 5.11. Figure 5.11a and b compare the open-loop and closed-loop (equipped with conventional ANN) area frequency response following a 0.1 pu step increase in the area load at 2 s. The parameter a_i of sigmoid unit functions in the applied ANN is fixed at $a_i = 1$. Figure 5.11c shows the system response for the same test scenario, using the proposed FNN controller. Figure 5.11b and c shows that the closed-loop performance for the FNN controller is much better than that for the conventional ANN with a fixed structure activation function. Finally, Figure 5.11d depicts the changing in power coming to the control area from all Gencos following the step disturbance. This figure shows that the power is initially coming from all Gencos to respond to the load increase, which will result in a frequency drop that is sensed by the speed governors of all machines through their primary frequency control loops. But after a few seconds and at steady state the additional power comes from the participating unit(s) in AGC only, and other Gencos do not contribute to the AGC task.

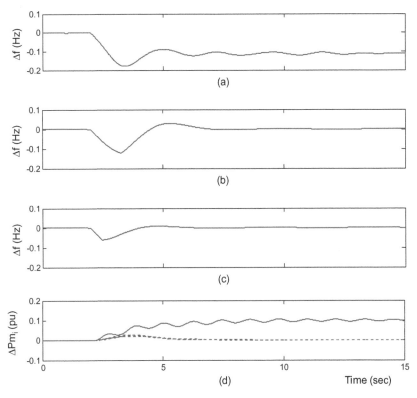

FIGURE 5.11
System response following 0.1 pu step load increase: (a) open-loop system, (b) conventional ANN, (c) proposed FNN, and (d) mechanical power changes (solid, Genco 1; dotted, other Gencos).

5.6.2 Three-Control Area

As another example, consider a power system with three control areas, as shown in Figure 5.12. Each control area has some Gencos with different parameters, and it is assumed that one generator unit with enough capacity is responsible for area frequency regulation (G11, G22, and G31 in areas 1, 2, and 3, respectively). Control area 1 delivers enough power from G11 and firm power from other Gencos to supply its load and support the AGC task. In case of a load disturbance, G11 must adjust its output to track the load changes and maintain the energy balance.

A control area may have a contract with a Genco in another control area. For example, control area 3 buys power from G11 in control area 1 to supply its load. The control areas are connected to the neighbor areas through L12, L13, and L23 interconnection lines. It is assumed that each Genco has one generator unit. The power system parameters are given in Bevrani et al.[40]

The proposed intelligent control design is applied to the three-control area power system described in Figure 5.12. Similar to the previous example, the

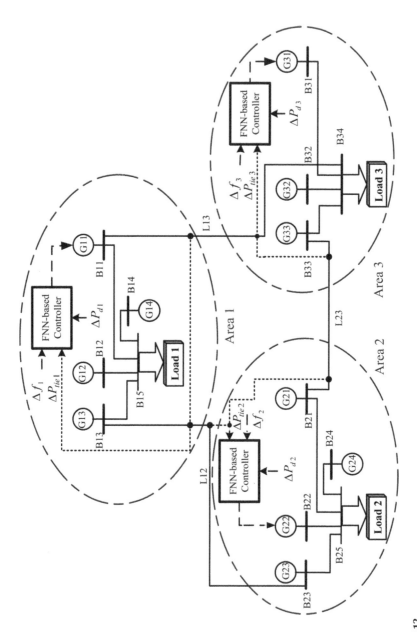

FIGURE 5.12
Three-control area power system.

TABLE 5.2

Learning Rates and Momentum Terms
for Proposed FNN

Control Area i	Learning Rates $[\eta_{1i}, \eta_{2i}]$	Momentum Terms $[\alpha_{1i}, \alpha_{2i}]$
1	[0.005, 0.002]	[0.07, 0.09]
2	[0.003, 0.001]	[0.05, 0.05]
3	[0.01, 0.002]	[0.01, 0.04]

initial connection weights and initial uniform random number of sigmoid function unit parameters for each control area are properly chosen. For instance, a set of suitable learning rates and momentum terms for area 1 are given in Table 5.2.

Figure 5.13a–c demonstrates a system response, following a 0.1 pu step load increase in each control area. In steady state, the frequency in each control area is properly returned to its nominal value. Comparing these results with the results obtained from ANN controllers with fixed structure, sigmoid functions for some test scenarios are given in Bevrani et al.[38] The comparison illustrates the effectiveness and capability of the proposed control design against the conventional NN-based AGC design.

Figure 5.13d demonstrates the disturbance rejection property of the closed-loop system. This figure shows the frequency deviation in all control areas, following a step disturbance (ΔP_d) of 0.01 pu on the interconnecting lines L12, L13, and L23 at $t = 2$ s.

Simultaneous learning of the connection weights and the sigmoid unit function parameters in the proposed method causes an increase in the number of adjustable parameters in comparison with the traditional method, and the proposed algorithm causes a reduction in the sensitivity of ANN to the parameters, such as connection weights, while increasing the sensitivity of ANN to the SFPs. However, in the proposed structure, the training of SFPs causes a change in the shape of individual sigmoid functions according to input space and reference signal and achieves betterment convergence and performance than traditional ANNs.

5.7 Summary

Power system operation and control took decades to shape, having been modified with increasing availability of new, powerful mathematical and computational tools. One of the modern tools is artificial neural networks

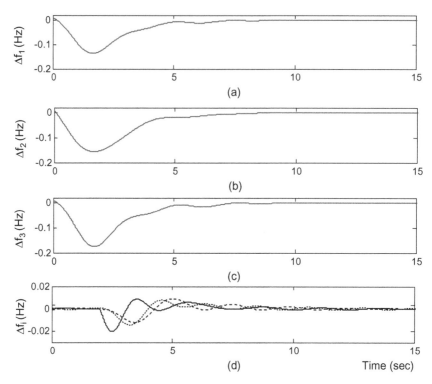

FIGURE 5.13
Frequency deviation in (a) area 1, (b) area 2, (c) area 3, following simultaneous 0.1 pu step load increases in three areas. (d) System response due to a 0.01 pu step disturbance on interconnecting lines at 2 s.

(ANNs). Over recent years, due to their ability to learn complex nonlinear functional relationships, ANNs have been suggested for many industrial systems using different configurations. At present, the methods provided by neural networks have matured and have been widely used in power system modeling, identification, and control.

In this chapter, a methodology for AGC design using flexible neural networks in a restructured power system has been proposed. Design strategy includes enough flexibility to set a desired level of performance. The proposed control methodology was applied to a single- and three-control area power system under a bilateral AGC scheme. It is recognized that the learning of both connection weights and SFPs increases the power of learning algorithms, in keeping with a high capability in the training process. Simulation results demonstrated the effectiveness of the methodology. It has been shown that the suggested FNN-based supplementary frequency controllers give better ACE minimization and a proper convergence to the desired trajectory than the traditional ANN ones.

References

1. H. Bevrani. 2009. *Robust power system frequency control.* New York: Springer.
2. H. Bevrani. 2004. Decentralized robust load-frequency control synthesis in restructured power systems. PhD dissertation, Osaka University.
3. H. L. Zeynelgil, A. Demiroren, N. S. Sengor. 2002. The application of ANN technique to automatic generation control for multi-area power system. *Elect. Power Energy Syst.* 24:345–54.
4. F. Beaufays, Y. Abdel-Magid, B. Widrow. 1994. Application of neural networks to load-frequency control in power systems. *Neural Networks* 7(1):183–94.
5. D. K. Chaturvedi, P. S. Satsangi, P. K. Kalra. 1999. Load frequency control—A generalized neural network approach. *Elect. Power Energy Syst.* 21:405–15.
6. Y. L. Karnavas, D. P. Papadopoulos. 2002. AGC for autonomous power system using combined intelligent techniques. *Elect. Power Syst. Res.* 62:225–39.
7. M. Djukanovic, M. Novicevic, D. J. Sobajic, Y. P. Pao. 1995. Conceptual development of optimal load-frequency control using artificial neural networks and fuzzy set theory. *Eng. Intelligent Syst. Elect. Eng. Commun.* 2:95–108.
8. H. Bevrani. 2002. A novel approach for power system load frequency controller design. In *Proceedings of IEEE/PES T&D 2002*, Asia Pacific, Yokohama, Japan, vol. 1, pp. 184–89.
9. A. Demiroren, N. S. Sengor, H. L. Zeynelgil. 2001. Automatic generation control by using ANN technique. *Elect. Power Components Syst.* 29:883–96.
10. T. P. I. Ahamed, P. S. N. Rao. 2006. A neural network based automatic generation controller design through reinforcement learning. *Int. J. Emerging Elect. Power Syst.* 6(1):1–31.
11. L. D. Douglas, T. A. Green, R. A. Kramer. 1994. New approaches to the AGC nonconforming load problem. *IEEE Trans. Power Syst.* 9(2):619–28.
12. H. Shayeghi, H. A. Shayanfar. 2006. Application of ANN technique based on Mu-synthesis to load frequency control of interconnected power system. *Elect. Power Energy Syst.* 28:503–11.
13. M. Teshnehlab, K. Watanabe. 1999. *Intelligent control based on flexible neural networks.* Dordrecht, NL: Kluwer Publishers.
14. W. W. McCulloch, W. Pitts. 1943. A logical calculus of ideas imminent in nervous activity. *Bull. Math. Biophys.* 5:115–33.
15. F. Rosenblatt. 1961. *Principles of neurodynamics.* Washington, DC: Spartan Press.
16. J. L. McClelland, D. E. Rumelhart. 1986. *Parallel distributed processing explorations in the microstructure of cognition: Psychological and biological models.* Vol. 2. Cambridge, MA: MIT Press.
17. A. Zilouchian, M. Jamshidi. 2001. *Intelligent control systems using soft computing methodologies.* Boca Raton, FL: CRC Press.
18. S. Haykin. 1994. *Neural networks.* New York: IEEE Press and Macmillan.
19. K. S. Narendra. 2003. Identification and control. In *The handbook of brain theory and neural networks*, ed. M. A. Arbib, 547–51. Cambridge, MA: MIT Press.
20. S. Haykin, ed. 2001. *Kalman filtering and neural networks.* New York: John Wiley & Sons.
21. M. M. Gupta, L. Jin, N. Homma. 2003. *Static and dynamic neural networks: From fundamentals to advanced theory.* New York: John Wiley & Sons.
22. F. Fogelman-Soulie, P. Gallinari, Y. LeCun, S. Thiria. 1987. Automata networks and artificial intelligence. In *Automata networks in computer science: Theory and applications,* 133–86. Princeton, NJ: Princeton University Press.

23. Y. LeCun. 1988. A theoretical framework for back-propagation. In *Proceedings 1988, Connectionist Model Summer School*, ed. D. Touretzky, C. Hinton, T. Sejnowski, 21–28. Pittsburgh, PA: Morgan Kaufmann.

24. F. L. Lewis, J. Campos, R. Selmic. 2002. *Neuro-fuzzy control of industrial systems with actuator nonlinearities*. Philadelphia: SIAM.

25. P. J. Werbos. 1991. A menu of designs for reinforcement learning over time. In *Neural networks for control*, ed. W. T. Miller, R. S. Sutton, P. J. Werbos, 67–95. Cambridge, MA: MIT Press.

26. J. Sarangapani. 2006. *Neural network control of nonlinear discrete-time systems*. Boca Raton, FL: CRC Press.

27. M. T. Rosenstein, A. G. Barto. 2004. Supervised actor-critic reinforcement learning. In *Handbook of learning and approximate dynamic programming*, ed. J. Si, A. G. Barto, W. B. Powell, D. Wunsch, 359–80. IEEE Press.

28. A. Pacut. 2003. Neural techniques in control. In *Neural networks for instrumentation, measurement and related industrial applications*, ed. S. Ablameyko et al., 79–117. Amsterdam, NL: IOS Press.

29. U. Halici, K. Leblebicioglu, C. Özegen, S. Tuncay. 2000. Recent advances in neural network applications in process control. In *Recent advances in artificial neural networks design and applications*, ed. L. Jain, A. M. Fanelli, 239–300. Boca Raton, FL: CRC Press.

30. A. Y. Alanis, E. N. Sanchez. 2009. Discrete-time reduced order neural observers. In *Advances in Computational Intelligence*, eds. W. Yu, E. N. Sanchez. AISC61: 113–22, Berlin Heidelberg: Springer-Verlag.

31. S. Ferrari. 2002. Algebraic and adaptive learning in neural control systems. PhD thesis, Princeton University.

32. I. Rivals, L. Personnaz. 2000. Nonlinear internal model control using neural networks: Applications to processes with delay and design issues. *IEEE Trans. Neural Networks* 11(1):80–90.

33. A. I. Galushkin. 2007. *Neural networks theory*. New York: Springer.

34. UCPTE. 1999. *UCPTE rules for the co-ordination of the accounting and the organization of the load-frequency control*. UCTE. 1999. UCTE Ground rules for the co-ordination of the accounting and the organization of the load-frequency control. Available: www.ucte.org.

35. R. D. Chritie, A. Bose. 1996. Load frequency control issues in power system operation after deregulation. *IEEE Trans. Power Syst.* 11(3):1191–200.

36. B. Delfino, F. Fornari, S. Massucco. 2002. Load-frequency control and inadvertent interchange evaluation in restructured power systems. *IEE Proc. Gener. Transm. Distrib.* 149(5):607–14.

37. G. Dellolio, M. Sforna, C. Bruno, M. Pozzi. 2005. A pluralistic LFC scheme for online resolution of power congestions between market zones. *IEEE Trans. Power Syst.* 20(4):2070–77.

38. H. Bevrani, T. Hiyama, Y. Mitani, K. Tsuji, M. Teshnehlab. 2006. Load-frequency regulation under a bilateral LFC scheme using flexible neural networks. *Eng. Intelligent Syst. J.* 14(2):109–17.

39. S. Haykin. 1999. *Neural networks: A comprehensive foundation*. 2nd ed. Upper Saddle River, NJ: Prentice Hall.

40. H. Bevrani, Y. Mitani, K. Tsuji. 2004. On robust load-frequency regulation in a restructured power system. *IEEJ Trans. Power Energy* 124(2):190–98.

6

AGC Systems Concerning Renewable Energy Sources

There is a rising interest in the impacts of renewable energy sources (RESs) on power system operation and control, as the use of RESs increases worldwide. A brief survey on the existing challenges and recent developments in this area is presented in Chapter 3. As mentioned, the renewable integration impacts may become more significant at a higher size of penetrations, and the range of impacts can be analyzed based on different system characteristics, penetration levels, and study methods.

The RESs certainly affect the dynamic behavior of the power system in a way that might be different from conventional generators. Conventional power plants mainly use synchronous generators that are able to continue operation during significant transient faults. If a large amount of wind generation is tripped because of a fault, the negative effect of that fault on the power system control and operation, including the AGC issue, could be magnified. High renewable energy penetration in power systems may increase uncertainties during abnormal operation, introduce several technical implications, and open important questions as to what happens to the AGC requirement in the case of adding numerous RESs to the existing generation portfolio, and whether the traditional power system control approaches to operation in the new environment are still adequate.

Integration of RESs into power system grids has impacts on optimum power flow, power quality, voltage and frequency control, system economics, and load dispatch. Regarding the nature of RES power variation, the impact on the AGC issue has attracted increasing research interest during the last decade.[1] Significant interconnection frequency deviations can cause under- or overfrequency relaying and disconnect some loads and generations. Under unfavorable conditions, this may result in a cascading failure and system collapse.[2]

This chapter covers the AGC system and related issues concerning the integration of new renewable power generation in power systems with a new perspective. The impact of power fluctuation produced by variable renewable sources (such as wind and solar units) on system frequency performance is presented. An updated power system frequency response model for AGC analysis considering RESs and associated issues is introduced. Some nonlinear time-domain simulations on the standard nine-bus and thirty-nine-bus test systems are presented to show that the simulated results agree with those predicted analytically. Emergency frequency control concerning RESs, particularly wind power, is discussed. Finally, the need for revising the frequency performance standards, further research, and new intelligent AGC schemes is emphasized.

6.1 An Updated AGC Frequency Response Model

When renewable power plants are introduced into the power system, an additional source of variation is added to the already variable nature of the system. To analyze the variations caused by RES units, the total effect is important, and every change in RES power output does not need to be matched one for one by a change in another generating unit moving in the opposite direction. Instantaneous fluctuations in load and RES power output might amplify each other, be completely unrelated to each other, or cancel each other out.[3] However, the slow RES power fluctuation dynamics and total average power variation negatively contribute to the power imbalance and frequency deviation, which should be taken into account in the AGC scheme. This power fluctuation must be included in the conventional AGC structure.[1]

A generalized AGC model in the presence of RES is shown in Figure 6.1. Here, for simplicity, the corresponding blocks for GRC, governor dead-band, and time delays are not included. To cover the variety of generation types in the control area, different values for turbine-governor parameters and the generator regulation parameters are considered. Figure 6.1 shows the block diagram of a typical control area with n generating units. The shown blocks and parameters are defined as follows: Δf is the frequency deviation, ΔP_m is the mechanical power, ΔP_C is the supplementary frequency control action, ΔP_L is the load disturbance, H_{Sys} is the equivalent inertia constant, D_{Sys} is the equivalent damping coefficient, β is the frequency bias, R_i is the drooping characteristic, ΔP_p is the primary frequency control action, α_i is the participation factor, ΔP_{RES} is the RES power fluctuation, ACE is the area control error, $M_i(s)$ is the governor-turbine model, and finally, $\Delta P'_L$ and $\Delta P'_{tie}$ are augmented local load change and tie-line power fluctuation signals, respectively.

The well-known conventional AGC frequency response model is discussed in Bevrani,[1] and briefly in Chapter 2. Following a load disturbance within a control area, the frequency of the area experiences a transient change and the feedback mechanism generates an appropriate rise or lower signal to the participating generator units according to their participation factors, to make the generation follow the load. In the steady state, the generation is matched with the load, driving the tie-line power and frequency deviations to zero. As there are many conventional generators in each area, the control signal has to be distributed among them in proportion to their participation. In

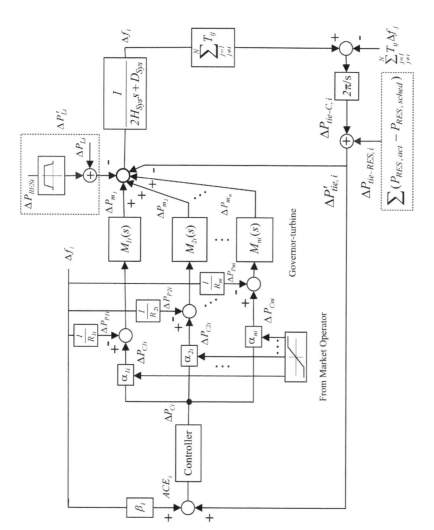

FIGURE 6.1
AGC model considering RES power fluctuation.

typical AGC implementations, the system frequency gradient and ACE signal must be filtered to remove noise effects before use. The ACE signal then is often applied to the controller block. The controller sends higher or lower pulses to the generating plants if its ACE signal exceeds a standard limit.

In the new AGC scheme, the updated ACE signal should represent the impacts of renewable power on the scheduled flow over the tie-line, as well as the local power fluctuation via the area frequency. The ACE signal is conventionally defined as a linear combination of frequency and tie-line power changes as follows:[1]

$$ACE = \beta \Delta f + \Delta P_{tie} \tag{6.1}$$

where, in a conventional power system, ΔP_{tie} (ΔP_{tie-C}) is the difference between the actual (*act*) and scheduled (*sched*) power flows over the tie-lines.

$$\Delta P_{tie-C} = \sum (P_{tie,act} - P_{tie,sched}) \tag{6.2}$$

The difference between the updated AGC model in Figure 6.1 and the conventional one (Figure 2.11) is in the following two new signals, representing the dynamic impacts of RESs on local load and tie-line power changes ($\Delta P'_L$, $\Delta P'_{tie-RES}$):

$$\Delta P'_L(s) = \Delta P_{RES}(s) - \Delta P_L(s) \tag{6.3}$$

$$\Delta P_{tie-RES} = \sum (P_{tie-RES,act} - P_{tie-RES,sched}) \tag{6.4}$$

In addition to the conventional power flow in power system tie-lines (ΔP_{tie-C}), for a considerable amount of renewable power, the transferred RES power through the tie-lines ($\Delta P_{tie-RES}$) should be considered. Therefore, the updated tie-line power deviation can be expressed as follows:

$$\Delta P'_{tie} = \Delta P_{tie-C} + \Delta P_{tie-RES}$$
$$= \sum (P_{tie-C,act} - P_{tie-C,sched}) + \sum (P_{tie-RES,act} - P_{tie-RES,estim}) \tag{6.5}$$

The total RES power flow change is usually smooth compared to variation impacts from the individual RES units. Using Equations 6.1 and 6.5, the updated ACE signal can be completed as Equation 6.6:

$$ACE = \beta \Delta f + \Delta P'_{tie}$$
$$= \beta \Delta f + \left(\sum (P_{tie-C,act} - P_{tie-C,sched}) + \sum (P_{tie-RES,act} - P_{tie-RES,sched}) \right) \tag{6.6}$$

where $P_{\text{tie-C,act}}$, $P_{\text{tie-C,sched}}$, $P_{\text{tie-RES,act}}$, and $P_{\text{tie-RES,sched}}$ are actual conventional tie-line power, scheduled conventional tie-line power, actual RES tie-line power, and scheduled RES tie-line power, respectively.

In managing the required regulated power, considering the rapid growth of variable renewable generation and the resulting impacts on power system performance is an important issue in a modern AGC system. For example, consider a power system with a high penetration of wind power. For a larger amount of fluctuating wind power, a greater amount of power reserve is needed to cover periods when there is no wind. On the other hand, managing surplus electricity during periods of strong wind could be also considered a challenge. Demand side control and intelligent power price management through intelligent meters and intelligent communications are already suggested as proper solutions for the mentioned challenge.[4] However, it may not be sufficient to rapidly increase RESs among the power system. The contribution of a renewable power plant (such as a wind farm) in AGC function via an intelligent configuration can be considered as an ultimate solution.

In combination with advanced forecasting techniques, it is now possible to design variable generators with a full range of performance capability that is comparable, and in some cases superior, to conventional synchronous generators. For example, modern wind generator control systems can provide an automatic response to frequency that is similar to the governor response on steam turbine generators. Unlike a typical thermal power plant whose output ramps downward rather slowly, wind farms can react quickly to a dispatch instruction, taking seconds, rather than minutes.

Many modern wind turbines are capable of pitch control, which allows their output to be modified (curtailed) in real time by adjusting the pitch of the turbine blades. By throttling back their output, wind plants are able to limit or regulate their power output to a set level or to set rates of change by controlling the power output on individual turbines.[5,6] This capability can be used to limit the ramp rate or power output of a wind generator, and it can also contribute to power system AGC. Some types of wind turbine generators are also capable of controlling their power output in real time in response to variations in grid frequency using variable speed drives.[7] Therefore, variable generation resources, such as wind power facilities, can be equipped to provide governing and participate in the AGC task as well as conventional generators. With the continued maturing of the technology, some renewable generators may participate in AGC systems in the future.

In the introduced generalized AGC model (Figure 6.1), it is assumed that only conventional generating units are participating in the AGC function. However, as mentioned above, RESs are needed to actively participate in the AGC issue and maintain system reliability along with conventional generation. In this case, the AGC model given in Figure 6.1 can be generalized as schematically shown in Figure 6.2. Here, renewable power plants $(RP_{1i}(s), \ldots, RP_{mi}(s))$, such as wind farms that can provide a considerable amount of power, also participate in the AGC system by producing regulation of renewable

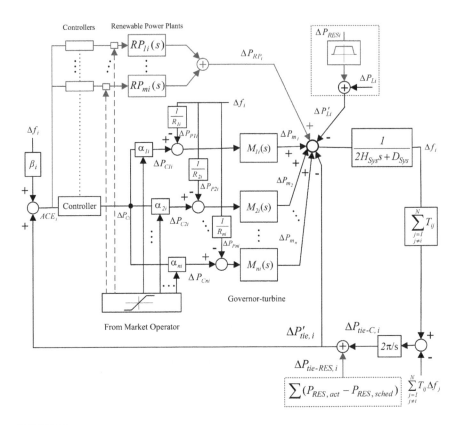

FIGURE 6.2
Generalized AGC model considering renewable power regulation.

power (ΔP_{RPi}). Similar to conventional generating units, the participation factor for each power plant is determined by a market operator. These power plants may use individual controllers, but those controllers should be coordinated with each other, as well as conventional ones. An intelligent coordinator block may provide a suitable solution for this issue in the future.

6.2 Frequency Response Analysis

Considering the effect of primary and supplementary frequency controls in Figure 6.1, the system frequency can be obtained as

$$\Delta f(s) = \frac{1}{2H_{Sys}s + D_{Sys}} \left[\sum_{k=1}^{n} \Delta P_{m_k}(s) - \Delta P_L'(s) - \Delta P_{tie}'(s) \right] \qquad (6.7)$$

where

$$\Delta P_{m_k}(s) = M_k(s)[\Delta P_{C_k}(s) - \Delta P_{P_k}(s)] \tag{6.8}$$

and

$$\Delta P_{P_k}(s) = \frac{\Delta f(s)}{R_k} \tag{6.9}$$

ΔP_P and ΔP_C are primary (governor natural response) and supplementary frequency control actions. Similar to today's power systems, it is assumed that the renewable power plants do not contribute to the load tracking and frequency regulation ($\Delta P_{RP} = 0$). Equations 6.8 and 6.9 can be substituted into Equation 6.7 with the result

$$\Delta f(s) = \frac{1}{2H_{Sys}s + D_{Sys}} \left(\sum_{k=1}^{n} M_k(s)[\Delta P_{C_k}(s) - \frac{1}{R_k}\Delta f(s)] - \Delta P_L'(s) - \Delta P_{tie}'(s) \right) \tag{6.10}$$

For the sake of load disturbance analysis it is usual to consider $\Delta P_L(s)$ in the form of a step function, i.e.,

$$\Delta P_L'(s) = \frac{\Delta P_L'}{s} \tag{6.11}$$

Substituting $\Delta P_L'(s)$ in Equation 6.11 and summarizing the result yields

$$\Delta f(s) = \frac{1}{g(s)} \left[\sum_{k=1}^{n} M_k(s)\Delta P_{C_k}(s) - \Delta P_{tie}'(s) \right] - \frac{1}{sg(s)}\Delta P_L' \tag{6.12}$$

where

$$g(s) = 2H_{Sys}s + D_{Sys} + \sum_{k=1}^{n} \frac{M_k(s)}{R_k} \tag{6.13}$$

Several low-order models for representing turbine-governor dynamics, $M_i(s)$, to use in power system frequency analysis and AGC design have been proposed. In these models, the slow system dynamics of the boiler and the fast generator dynamics are ignored. A second-order model was introduced in Elgerd and Fosha.[8] Also, a simplified first-order turbine-governor model was proposed.[9]

Substituting Mi(s) from Elgerd and Fosha[8] or Anderson and Mirheydar[9] in Equations 6.12 and 6.13, and using the final value theorem, the frequency deviation in the steady state can be obtained from Equation 6.12 as follows:

$$\Delta f_{ss} = \underset{s \to 0}{Lim}\, s.\Delta f(s) = \frac{1}{g(0)}[\Delta P_C - \Delta P'_{tie}] - \frac{1}{g(0)}\Delta P'_L \qquad (6.14)$$

where

$$\Delta P_C = \underset{s \to 0}{Lim}\, s.\sum_{k=1}^{n} M_k(s)\Delta P_{C_k}(s) \qquad (6.15)$$

$$\Delta P'_{tie} = \underset{s \to 0}{Lim}\, s.\Delta P'_{tie}(s) \qquad (6.16)$$

$$g(0) = D_{Sys} + \sum_{k=1}^{n} \frac{1}{R_k} = D_{Sys} + \frac{1}{R_{Sys}} \qquad (6.17)$$

Here, R_{Sys} is the equivalent system droop characteristic, and

$$\frac{1}{R_{Sys}} = \sum_{k=1}^{n} \frac{1}{R_k} \qquad (6.18)$$

By definition,[10] $g(0)$ is equivalent to the system's frequency response characteristic (β).

$$\beta = D_{Sys} + \frac{1}{R_{Sys}} \qquad (6.19)$$

Using Equation 6.17, Equation 6.14 can be rewritten into the following form:

$$\Delta f_{ss} = \frac{\Delta P_C - \Delta P'_{tie} - \Delta P'_L}{D_{Sys} + 1/R_{Sys}} \qquad (6.20)$$

Equation 6.20 shows that if the disturbance magnitude matches the available power reserve (supplementary control), $\Delta P_C = \Delta P'_{tie} + \Delta P'_L$, the frequency deviation converges to zero in the steady state. Since the value of a droop characteristic R_k is bounded between about 0.05 and 0.1 for most generator units ($0.05 \le R_k \le 0.1$),[9] for a given control system, according to Equation 6.18, we can write $R_{Sys} \le R_{min}$. For a small $D_{Sys} R_{Sys}$, Equation 6.20 can be reduced to

$$\Delta f_{ss} = \frac{R_{Sys}(\Delta P_C - \Delta P'_{tie} - \Delta P'_L)}{(D_{Sys} R_{Sys} + 1)} \cong R_{Sys}(\Delta P_C - \Delta P'_{tie} - \Delta P'_L) \qquad (6.21)$$

Without a supplementary frequency control signal ($\Delta P_C = 0$), the steady-state frequency deviation will be proportional to the disturbance magnitude as follows:

$$\Delta f_{ss} = -\frac{R_{Sys}(\Delta P'_{tie} + \Delta P'_L)}{(D_{Sys}R_{Sys} + 1)} \tag{6.22}$$

6.3 Simulation Study

The power outputs of some RESs, such as solar and wind power generation systems, are dependent on weather conditions, seasons, and geographical location. Therefore, they can significantly influence the system frequency regulation performance. This section provides a simulation study on the impacts of solar and wind power units on the power system frequency. The simulation study is performed on two standard test systems: nine-bus, three generating units; and thirty-nine-bus, eleven generators.

6.3.1 Nine-Bus Test System

The first case study, as shown in Figure 6.3, is a nine-bus power system including 567.5 MW of conventional generation, a 2,000 kW photovoltaic (PV) unit, and two wind farms with 35 MW. This test system is considered to simulate the impact of existing RESs (PV and wind turbine units) on the system frequency performance. It is assumed that generator G_2 is responsible for regulating system frequency using a simple proportional-integral (PI) controller. The nine-bus system parameters are assumed to be the same as those used in Anderson and Fouad.[11]

For the sake of simulation, random variations of solar isolation and wind velocity have been taken into account. A combination of variable and fixed wind turbines has been used on wind farms. The variation of produced powers by wind farms and PV sources is the source of frequency variation in the study system. The wind velocities, V_W (m/s), the output power of wind farms, P_{WT} (MW), and the output power of the PV unit, P_{PV} (MW), are shown in Figure 6.4. The system response is shown in Figure 6.5. This figure shows the produced power by conventional generators, P_G (MW), and the frequency variation, Δf (Hz), at generator terminals.

The system response following connection of the low-capacity PV unit (only) is shown in Figure 7.5b (dashed line), as well as the frequency variation in the presence of both PV and wind turbine units (solid line). When wind power is a part of the power system, additional imbalance is created when the actual wind output deviates from its forecast.

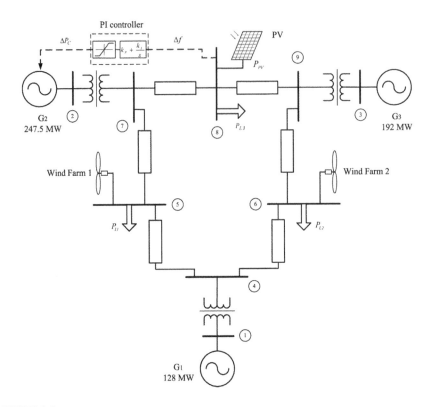

FIGURE 6.3
Nine-bus system: three generators, two wind farms, and one PV unit.

Fast movements in wind power output are combined with fast movements in load and other resources. Scheduling conventional generator units to follow load (based on the forecasts) may also be affected by wind power output. Errors in load forecasts are generally uncorrelated with errors in wind forecasts. The initial frequency rate change for the given simulation example, following a 0.05 pu step load disturbance, is shown in Figure 6.6. A similar test is repeated on the nine-bus system without RES units, and a larger frequency rate change is achieved.

As shown in Bevrani,[1] the frequency gradient in a power system is proportional to the magnitude of total load-generation imbalance. The factor of proportionality is the system inertia. In fact, the inertia constant is loosely defined by the mass of all the synchronous rotating generators and motors connected to the system. For a specific load decrease, if H is high (e.g., due to integration of wind turbines), then the frequency will fall slowly, and if H is low, then the frequency will fall faster.

Since the system inertia determines the sensitivity of the overall system frequency, it plays an important role in the frequency regulation issue. A large interconnected power system generally has sizable system inertia, and

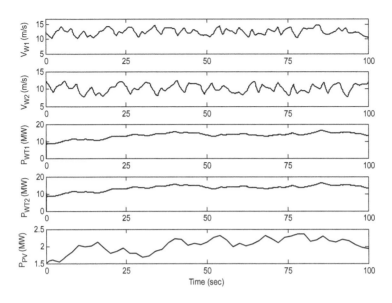

FIGURE 6.4
Wind velocity in the wind farms, and output power of RES units.

frequency deviation in the presence of wind and solar power variations is small. In other words, a larger electricity industry may be more capable of absorbing variations in electricity output from RESs. However, the combination of RES systems with system inertia of a small isolated power system must be considered, and the AGC designs need to consider altering their frequency control strategies to avoid long rates of change of the system frequency.

6.3.2 Thirty-Nine-Bus Test System

A network with the same topology as the well-known IEEE thirty-nine-bus test system (Figure 6.7) is considered as another study system to simulate the impact of RESs on the system frequency performance. The test system has ten generators, nineteen loads, thirty-four transmission lines, and twelve transformers. Here, the test system is updated by two wind farms in areas 1 and 3, and a photovoltaic (PV) unit in area 2. The total generation includes 842 MW of conventional power, 2,000 kW of solar power, and 46 MW of wind power. The test system is organized into three areas. The amounts of load in areas 1, 2, and 3 are 265.5, 233, and 125 MW, respectively.

All power plants in the power system are equipped with a speed governor and power system stabilizer (PSS). However, only one generator in each area is responsible for the AGC task using a PI controller: G1 in area 1, G9 in area 2, and G4 in area 3. The simulation parameters for the generators, loads, lines, and transformers of the test system are assumed to be the same as those given in Bevrani et al.[12]

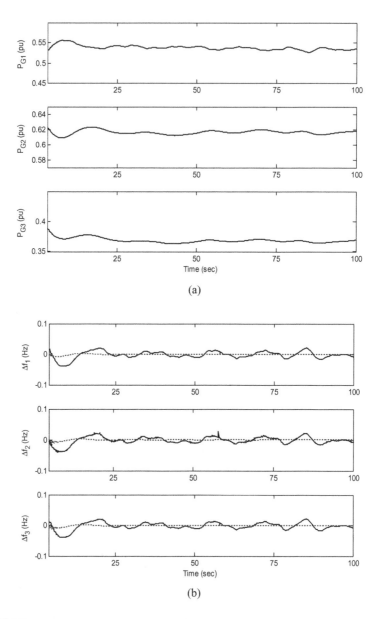

(a)

(b)

FIGURE 6.5

Conventional generator response: (a) output power and (b) frequency change (dashed line shows response with PV unit only).

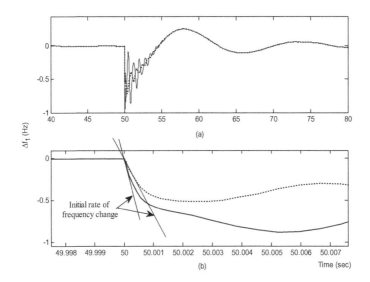

FIGURE 6.6
(a) Frequency deviation following a 0.05 pu step load disturbance at 50 s, with (dotted) and without (solid) RES units. (b) A zoomed view around 50 s.

A combination of fixed speed and double-fed induction wind turbine generators (WTGs) is used on the wind farms. Both WTG types cause variations in output electric power. For fixed speed wind turbines, variations in wind speed cause this variation. The produced power by double-fed induction generators (DFIGs) also varies with wind speed, although the torque-speed controller provides that this variability is less volatile than that for the fixed speed turbines. For a single turbine, this variability depends on time interval, location, and terrain features.[13]

Here, dynamics of WTGs, including the pitch angle control of the blades, are also considered. The start-up and rated wind velocities for the wind farms are specified at about 8.16 and 14 m/s, respectively. Furthermore, the pitch angle controls for the wind blades are activated only beyond the rated wind velocity. The pitch angles are fixed to 0° at the lower wind velocity below the rated one. The wind velocity, V_{Wind} (m/s), the total output power of wind farms, P_{WT} (MW), and the output power of the PV unit, P_{PV} (MW), are shown in Figure 6.8a–c, respectively.

The corresponding overall power system frequency deviation due to wind power fluctuations is also shown in Figure 6.8d. The power imbalance may lead to frequency deviations from the nominal value (60 Hz in the present example). From the power quality point of view, frequency deviations should be limited in a specified standard band. As shown in the simulation results, an averaged/filtered form of wind (and solar) power variations is reflected in the overall system frequency.

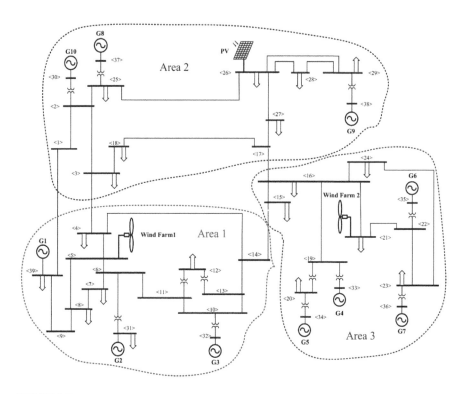

FIGURE 6.7
Single-line diagram of thirty-nine-bus test system.

As discussed in Section 6.4, adding of RES units (specifically wind generators) to a power system leads to an increase in total system inertia. The most pronounced effect of high values of inertia is to reduce the initial rate of frequency decline, and to reduce the maximum deviation. To explore the above issue, system response following a step load disturbance was investigated. A step load increase is considered at 10 s in each area as follows: 3.8% of area load at bus 8 in area 1, 4.3% of area load at bus 3 in area 2, and 6.4% of area load at bus 16 in area 3 have been changed. The applied step load disturbances, system frequency, and frequency gradient of the system with and without RESs are shown in Figure 6.9a–c, respectively. For having a clear comparison, a zoomed view of the rate of frequency changes of around 10 s is also shown in Figure 6.9d.

It can be seen that the frequency drop has reduced. It is justifiable by considering the larger inertia due to the adding of large wind farms to the system. The higher inertia value results in a lower drop in frequency.[10] But when renewable energy sources replace large conventional generating units, the inertia of the power system often falls and may increase variation in system frequency.

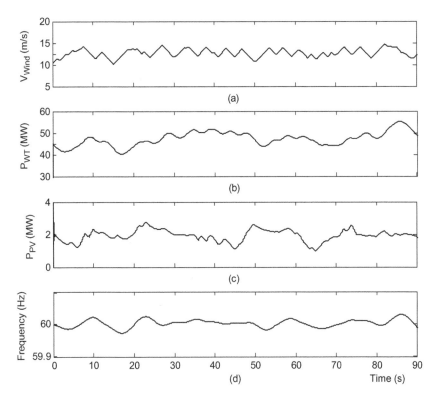

FIGURE 6.8
System response: (a) wind velocity, (b) wind power, (c) PV power, and (d) system frequency.

The lack of power system inertia due to replacing the conventional power plants by RES units is particularly relevant for the stability of island power systems with large amounts of renewable power. But large-scale renewable power in areas of large interconnected systems has also been shown to influence frequency and power oscillations.[14] Particularly, this issue will be important concerning the increase of the amount of installed wind power.

Fixed speed wind turbines with directly connected generators increase the system inertia, but the inertia is typically less than that in conventional generators of the same capacity. Standard variable speed wind turbines do not contribute to the power system inertia, because their rotational speeds are independent of frequency; neither does solar power (PV), simply because there is no rotating mass to provide the inertia. Due to the relatively low share of installed PV capacity, this is not likely to be an issue in the near future.[14]

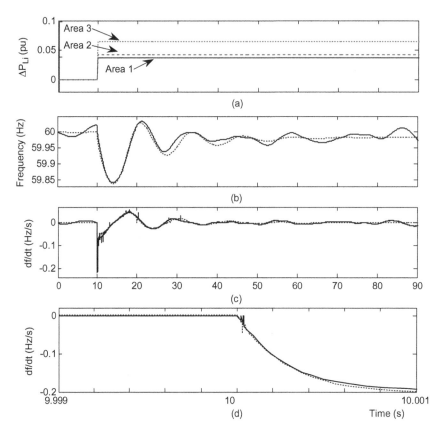

FIGURE 6.9
System response following simultaneous load disturbances with (solid) and without (dotted) RESs: (a) load disturbances, (b) system frequency, (c) rate of frequency change, and (d) zoomed view of the rate of frequency changes around 10 s.

6.4 Emergency Frequency Control and RESs

There are few reports on the role of distributed RESs in emergency conditions. The impact of distributed utilities on transmission stability is addressed in Donnelly et al.,[15] and an optimal load shedding strategy for power systems with distributed sources is introduced in Xu and Girgis.[16] The need for retuning of automatic under-frequency load shedding (UFLS), df/dt, relays has been emphasized.[1,17] The system performance and frequency stability during a severe short circuit and after a sudden loss of generation are discussed in Erlich and Shewarega.[18]

IEEE 1547[19] considers a small RES unit having less impact on system operation, but a large RES unit can have an impact on distribution system safety. This requirement is taken into account by allowing the network operator to specify the frequency setting and time delay for under-frequency trips. When frequency is out of the given protection range,[20] the RES unit shall cease to energize the area power system within the clearing time, as indicated. The clearing time is the period elapsed between the start of the abnormal condition and the RES ceasing to energize the power system.

A variety of studies are recommended to analyze the protection-based penetration limits with consideration of the RESs' capacity, location, and technology. The studies aid in determining mitigation strategies to increase the protection-based penetration limit. The loss of coordination, desensitization, nuisance fuse blowing, bidirectional relay requirements, and overvoltage should be studied in order to arrive at the penetration limits of RESs in an existing distribution system.[21,22]

To study the frequency behavior in a power system in response to the serious disturbances, and impacts of different types of wind turbines, the nine-bus power system example given in Figure 6.3 is examined. Figure 6.10a demonstrates the system response following outage of lines 8-7 and 8-9. Although the system remains stable, the steady-state frequency following the faults is a little bit changed from the nominal frequency. It is noteworthy that the test system is not equipped with an AGC system.

As a second test scenario, generator G2, which is the largest one in the test system, is tripped at t = 10 s, and the results are monitored for the following cases: without wind turbine, with 10% DFIG type penetration, with 10% induction generator (IG) type penetration, and with 10% IG type wind turbine compensated with a static compensator (STATCOM). Figure 6.10b shows the frequency response following this disturbance. The rate of frequency change is also illustrated in Figure 6.10c. These results are explained in Bevrani and Tikdari.[23] All four cases are unstable. Therefore, to reestablish the system frequency, one may use an UFLS scheme, as described in Chapter 2. The UFLS scheme sheds a portion of load demand when the frequency decline reaches the predetermined thresholds.

Recent studies show that to design an effective load shedding plan, considering both frequency and voltage indices is needed. In this direction, an intelligent-based power system emergency control has been proposed.[23] The developed scheme is summarized in Figure 6.11. A severe contingency triggers the proposed load shedding algorithm considering both voltage and frequency. For this purpose, a trained artificial neural network (ANN) uses the measured tie-line active/reactive powers to estimate the system P-V curve. The amount of load that should be shed is immediately computed using an estimated P-V curve.

The proposed ANN is a three-layer backpropagation neural network. The activation functions for the hidden layers are in the form of a tangent-sigmoid function, and the output activation functions are linear. The ANN first should be trained to predict the system P-V curve. The inputs of ANN are the severe contingency firing command, and the tie-line active and reactive powers before the related event.[23,24]

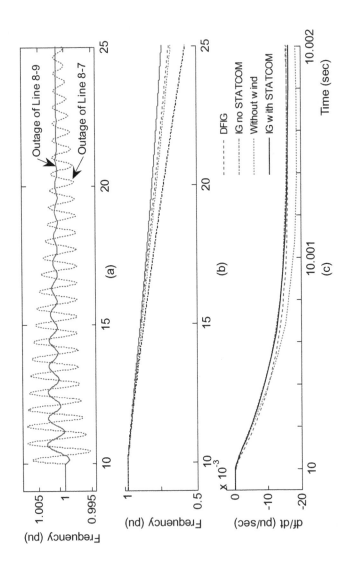

FIGURE 6.10
Nine-bus system response following serious disturbances: (a) line outage, (b) loss of G2, and (c) rate of frequency change for loss of G2.

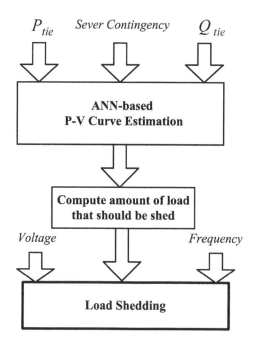

FIGURE 6.11
An intelligent-based emergency control scheme.

The effect of adding RES units to the distribution feeder can produce blind zones for protection devices or upset the coordination between two (or more) protective devices, and should be studied carefully.[25] In normal operation, protection devices are coordinated such that the primary protection operates before the backup can take action. Interconnecting distributed RESs increases the short-circuit level. Depending on the original protection coordination settings, along with the size, location, and type of the RESs, uncoordinated situations may be found. In these situations, the backup operates before the primary, which results in nuisance tripping to some of the loads.

The introduction of distributed RESs that can operate in islanding mode results in a complex problem that requires study to determine the necessary settings and changes needed for proper island operation. When parallel operation is lost, the RESs must separate themselves from the utility system quickly to support the substation reclose attempt. Detecting the loss of parallel operation (unintentional island formation) is done by establishing an over- or underfrequency (and over- or undervoltage) within which the distributed RES is allowed to operate. Under most circumstances, the frequency will quickly move outside of normal operation when parallel operation is lost.

Curtailing the megawatt (MW) output of wind generation is another method to restore the system frequency in emergency conditions. As wind penetration levels increase, the amount of curtailment and frequency of curtailment will increase. When wind generation levels are a high percentage of system demand, when the output from wind generators will differ significantly from the forecast output because of sudden unpredicted changes in weather patterns, and in case of simultaneous loss of a number of transmission lines, it may be necessary to curtail wind generation in order to manage the power system.[26]

6.5 Key Issues and New Perspectives

6.5.1 Need for Revision of Performance Standards

Deploying various types of RESs to take advantage of complementary patterns of production, locating variable resources across a large geographical region to leverage any fuel diversity that may exist, advanced control technology designed to address ramping, supply surplus conditions, and AGC show significant promise in managing variable generation characteristics. As mentioned in Chapter 2, the AGC system resides in the system control center and monitors the imbalance between generation and demand within a control area. In response to higher levels of variable generation, the AGC algorithms and parameters may need to be modified for better performance.

Within a control area, AGC adjusts supply automatically between dispatch intervals to ensure that the control area is contributing to maintain system frequency and keeps its interchange(s) with neighboring control area(s) at scheduled value(s).[7] Interconnection procedures and standards should be reviewed to ensure that the operating AGC scheme and its responses are in a consistent manner to all power generation technologies, including variable generation technologies. Variable generation technologies generally refer to generating technologies whose primary energy source varies over time and cannot reasonably be stored to address such variation. Uncertainty and variability are two major attributes of a variable generator that distinguish it from conventional forms of generation and may impact the overall system planning and operations.

Reliable power system operation requires an ongoing balancing of supply and demand in accordance with the prevailing AGC operating criteria and standards, such as those established by the North American Electric Reliability Council (NERC) and Union for the Coordination of Transmission of Electricity (UCTE). Power system operation is always in a changing state due to integration of new power sources, maintenance schedules, unexpected outages, and changing interconnection schedules and fluctuations in demand, generation, and power flow over transmission lines. The characteristics of the new installed power system equipment, their controls, and the actions of system operators play a significant role in ensuring that the AGC system in a bulk power system performs acceptably after disturbances and can be restored to a balanced state of power flow and frequency.[7]

As mentioned, the increasing share of renewable energy, which is difficult to predict accurately, may have an adverse impact on frequency quality. The existing frequency operating standards[17] need to change to allow for the introduction of renewable power generation, and allow for modern distributed generator technologies. It has been shown that the slow component of renewable power fluctuation negatively affects the performance standards, such as policy P1 of the UCTE performance standard,[27] or the control performance standards CPS1 and CPS2 introduced by NERC.[29] Therefore, the AGC issue and related standards may evolve into new guidelines.

Until now, wind turbine compatibility to the various standard requirements has been established only through specific tests or simulations that were performed by the manufacturers or other independent laboratories, upon demand of system operators. Standardized type tests have not been developed yet, due to the diversity of requirements appearing in grid codes,[30] performance standards, and the relatively limited time they have been in force. Moreover, testing the actual behavior of wind turbines during system faults presents significant difficulties, since on-site tests on installed machines are necessary, which will involve power system frequency and power fluctuations.

The standards redesign must be done in both normal and abnormal conditions, and should take account of operational experience on the initial frequency control schemes, and again used measurement signals, including tie-line power, frequency, and rate of frequency change settings. The new set of frequency performance standards is under development in many countries.[31] The new standards introduce the updated high- and low-trigger, abnormal, and relay limits applied to the interconnection frequency excursions. For high wind penetration, not only do frequency relay settings need revision, but current and voltage relays also need to be coordinated.[17,32] Protection schemes for distribution and transmission networks are one of the main problems posed by RESs in power systems. Changes in operational conditions and dynamic characteristics influence the requirements for protection parameters.[23]

The performance standards revision has already commenced in many countries.[1,32] In Australia, the Australian Electricity Market Commission (AEMC) has proposed revised technical rules for generator connection, including wind generators.[33] As well as meeting technical standards, generators are required to provide information on energy production via the system operator's supervisory control and data acquisition (SCADA) system. The National Electricity Market Management Company (NEMMCO) sets out a functional requirement for an Australian Wind Energy Forecasting System (AWEFS) for wind farms in market regions. In the United States, NERC is working to revise the conventional control performance standards.[31] The existing market rules and priority rules for the transport of RES electricity are also under reexamination by UCTE in Europe.[33]

6.5.2 Further Research Needs

Research in this area has received increasing attention. However, many issues are still in the early stages of development. Continued work is needed to design effective compensation methods. Additional research is required in understanding how future AGC systems should be designed to simplify the integration of RESs and other distributed generators in a highly competitive environment. Some important research needs for the future can be summarized as follows:

1. *Modeling, and aggregation techniques.* The rise in the share of RES production in the power system network is increasingly requiring an analysis of the system dynamic behavior of some incidents that may occur through an effective modeling. In real-time AGC operation, the characteristics of the overall power system must be understood to ensure reliable operation. For example, regulating reserves and ramping capabilities are critical attributes necessary to deal with the short-term uncertainty of demand and generation, as well as with the uncertainty in the demand forecasts, generation availability, and economical issues.

   The conventional power system is generally planned assuming the AGC system is functioning properly. However, a comprehensive approach is needed for planning from the AGC system through to the power system, particularly with the increased penetration of RESs' power on different control areas, which may severely stress the AGC operation and can also impact modern power system reliability. Therefore, these impacts need to be understood and resolved in the power system operation and control.

2. *Grid codes.* The ability to tolerate grid faults has become a key issue in the large-scale use of renewable power. This is usually reflected in the grid codes, the rules that govern the behavior of generating equipment connected to the grid. Further research is needed to define new grid codes for integration and participation of RESs in the AGC system and other ancillary services. Every country planning to develop large-scale renewable power now has grid codes dealing specifically with renewable power generation technologies. The main purpose is to ensure that renewable generators are able to stay connected to the grid during and after a grid fault such as a short circuit. If this does not happen, the resulting sudden loss of renewable generation can turn a minor grid fault into a more serious event.[14]

Grid connection guidelines are still a major controversial subject concerning the distributed RESs. The connection rules and technical requirements that differ from region to region make it all even more complicated. In order to allow a flexible and efficient introduction of RESs, there is a need for a single document being a consensus standard on technical requirements for RES interconnection rather than having the manufacturers and operators conform to numerous local practices and guidelines.[20]

3. *Intelligent system infrastructures and intelligent AGC schemes.* The change from a few large power plants to a much higher number of smaller plants implies a need to change the control system paradigm of the power system, from conventional to the intelligent side. The requirements to the control and communication infrastructure for integrating RESs for frequency regulation are very different from those of the existing AGC system, but this has to be a part of the total control concept of the new AGC structure that includes large central thermal generating units and large renewable power plants (such as wind farms). The infrastructure of the future intelligent AGC system should support the provision of ancillary services from sources other than the central large power plants. This is because the economic operation of future power systems will reduce the capacity of central power plants to remain online.

   More flexible and intelligent system infrastructures are required to facilitate a substantially higher degree of deregulation and higher amounts of renewable energy compared to today's power systems. The core of an intelligent system should include fast communication between the SCADA/AGC center and generating units, as well as between energy producers and energy consumers. Such communication might well be based on real-time pricing. In this direction, communication standards are important to ensure that the devices connected to the intelligent power system are compatible, and the ability of the SCADA and AGC systems to cover both scalability (large numbers of units) and flexibility (new types of units).

4. *New AGC participants and coordination.* One of main challenges for future power systems is to ensure that distributed RESs and decentralized generators provide the regulation services for which we currently rely on large central generating units. A future power system with many intelligent components capable of participating in the AGC issue could provide a desirable performance supporting a high proportion of renewable energy.

   Ancillary services provided by DGs/RESs in coordination with conventional large generating units will become increasingly important as the penetration of DGs/RESs increases. Therefore, managing the ramp rates of renewable power plants can be challenging for the

AGC system and market operator, particularly if down ramps occur as demand increases, and vice versa. Insufficient ramping and dispatchable capability on the remainder of the bulk power system can exacerbate these challenges. Ramping control could be as simple as electrically tripping all or a portion of the variable generation plant. However, more modern variable generation technologies allow for continuous dispatch of their output. Continuous ramp rate limiting and power limiting features are readily available for some renewable power generators. Modern control areas require power management on renewable power facilities, such that the market operator can reduce the ramp rate limit to a reliable limit that can be accommodated on the power system at that time.[7]

6.6 Summary

Reliable power system operation requires an ongoing balancing of supply and demand in accordance with established operating criteria. The AGC system provides for the minute-to-minute reliable operation of the power system by continuously matching the supply of electricity with the demand, while also ensuring the availability of sufficient supply capacity in future hours. To date, due to the integration of RESs and DGs, the AGC design has evolved into new guidelines. High-level integration of variable generation typically has not appreciably impacted the AGC performance and the system reliability.

This chapter presents an overview of the key issues concerning the integration of RESs into the power system frequency regulation that are of most interest today. The most important issues with the recent achievements in this literature are briefly reviewed. The impact of RESs on the frequency control problem is described. An updated AGC model is introduced. Power system frequency response in the presence of RESs and associated issues is analyzed, and the need for the revising of frequency performance standards is emphasized. Finally, a nonlinear time-domain simulation study for two power system examples is presented.

References

1. H. Bevrani. 2009. *Robust power system frequency control.* New York: Springer.
2. Y. V. Makarov, V. I. Reshetov, V. A. Stroev, et al. 2005. Blackout prevention in the United States, Europe and Russia. *Proc. IEEE* 93(11):1942–55.

3. R. Gross, P. Heptonstall, M. Leach, et al. 2007. Renewable and the grid: Understanding intermittency. *Energy* 160:31–41.
4. P. E. Morthorst, D. Vinther. 2009. Challenges for a future Danish intelligent energy system. In *The intelligent energy system infrastructure for the future*, ed. H. Larsen, L. S. Petersen, 15–20. RisØ Energy Report, vol. 8. Roskilde, Denmark: National Laboratory for Sustainable Energy.
5. J. Morel, H. Bevrani, T. Ishii, T. Hiyama. 2010. A robust control approach for primary frequency regulation through variable speed wind turbines. *IEEJ Trans. Power Syst. Energy.* 130(11): 1002–9.
6. L. Y. Pao, K. E. Johnson. 2009. A tutorial on the dynamics and control of wind turbines and wind farms. In *Proceedings of American Control Conference*, St. Louis, pp. 2076–89.
7. NERC. 2009. *Accommodating high levels of variable generation.* Special report. http://www.nerc.com/files/IVGTF_Report_041609.pdf (accessed May 17, 2010).
8. O. I. Elgerd, C. Fosha. 1970. Optimum megawatt-frequency control of multiarea electric energy systems. *IEEE Trans. Power Apparatus Syst.* PAS-89(4):556–63.
9. P. M. Anderson, M. Mirheydar. 1990. A low-order system frequency response model. *IEEE Trans. Power Syst.* 5(3):720–29.
10. P. Kundur. 1994. *Power system stability and control.* Englewood Cliffs, NJ: McGraw-Hill.
11. P. M. Anderson, A. A. Fouad. 1994. *Power system control and stability.* New York: IEEE Press.
12. H. Bevrani, F. Daneshfar, P. R. Daneshmand. 2010. Intelligent power system frequency regulation concerning the integration of wind power units. In *Wind power systems: Applications of computational intelligence*, ed. L. F. Wang, C. Singh, A. Kusiak, 407–37. Springer Book Series on Green Energy and Technology. Heidelberg: Springer-Verlag.
13. Power Systems Engineering Research Center (PSERC). 2009. *Impact of increased DFIG wind penetration on power systems and markets.* Final project report. PSERC.
14. P. Sørensen. 2009. Flexibility, stability and security of energy supply. In *The intelligent energy system infrastructure for the future*, ed. H. Larsen, L. S. Petersen, 25–29. RisØ Energy Report, vol. 8. Roskilde, Denmark: National Laboratory for Sustainable Energy.
15. M. K. Donnelly, J. E. Dagle, D. J. Trudnowski, et al. 1996. Impacts of the distributed utility on transmission system stability. *IEEE Trans. Power Syst.* 11(2):741–47.
16. D. Xu, A. A. Girgis. 2001. Optimal load shedding strategy in power systems with distributed generation. In *Proceedings of IEEE PES Winter Meeting*, vol. 2, Columbus, OH, USA. pp. 788–93.
17. H. Bevrani, G. Ledwich, J. J. Ford. 2009. On the use of df/dt in power system emergency control. Paper presented at Proceedings of IEEE Power Systems Conference & Exposition, Seattle, Washington.
18. I. Erlich, F. Shewarega. 2007. Insert impact of large-scale wind power generation on the dynamic behaviour of interconnected systems. Paper presented at Proceedings of iREP Symposium—Bulk Power System Dynamics and Control, Charleston, SC.

19. IEEE. 2003. *Standard for interconnection distributed resources with electric power system*. IEEE 1547. IEEE.
20. V. V. Thong, J. Driesen, R. Belmans. 2007. Overview and comparisons of existing DG interconnection standards and technical guidelines. In *Proceedings of International Conference on Clean Electrical Power—ICCEP*, Capri, IT, pp. 51–54.
21. A. T. Moore. 2008. *Distributed generation (DG) protection overview*. Technical report, University of Western Ontario. http://www.eng.uwo.ca/people/tsidhu/Documents/DG%20Protection%20V4.pdf.
22. T. Abel-Galil, A. Abu-Elanien, E. El-Saadany, et al. 2007. *Protection coordination planning with distributed generation*. Varennes, Quebec, CN: Natural Resources Canada—CETC.
23. H. Bevrani, A. G. Tikdari. 2010. An ANN-based power system emergency control scheme in the presence of high wind power penetration. In *Wind power systems: Applications of computational intelligence*, ed. L. F. Wang, C. Singh, A. Kusiak, 215–54. Springer Book Series on Green Energy and Technology. Heidelberg: Springer-Verlag.
24. H. Bevrani, A. G. Tikdari, T. Hiyama. 2010. An intelligent based power system load shedding design using voltage and frequency information. In *Proceedings of the International Conference on Modelling, Identification and Control*, Okayama, Japan, pp. 545–49.
25. C. Kwok, A. Morched. 2006. *Effect of adding distributed generation to distribution networks case study 3: Protection coordination considerations with inverter and machine based DG*. Varennes, Quebec, CN: Natural Resources Canada—CETC.
26. ESB National Grid. 2009. Options for operational rules to curtail wind generation. www.cer.ie/cerdocs/cer04247.doc.
27. UCTE. 2004. UCTE appendix to policy P1: Load-frequency control and performance. In *UCTE operation handbook*. Appendix 1: 1–27. Available www. ucte.org.
28. H. Banakar, C. Luo, B. T. Ooi. 2008. Impacts of wind power minute to minute variation on power system operation. *IEEE Trans. Power Syst.* 23(1):150–60.
29. M. Tsili, S. Papathanassiou. 2009. A review of grid code technical requirements for wind farms. *IET Renew. Power Gener.* 3(3):308–32.
30. NERC. 2007. *Balance resources and demand standard*. Ver. 2. http://www.nerc.com/~filez/standards/Balance-Resources-Demand.html.
31. H. Bevrani, A. Ghosh, G. Ledwich. 2010. Renewable energy sources and frequency regulation: Survey and new perspectives. *IET Renewable Power Gener.*, 4(5): 438–57.
32. AEMC. 2006. *Draft national electricity amendment (technical standards for wind generation and other generator connections)*. Rule 2006. http://www.aemc.gov.au.
33. EWIS. 2007. *Towards a successful integration of wind power into European electricity grids*. Final report. http://www.ornl.gov/~webworks/cppr/y2001/rpt/122302.pdf.

7

AGC Design Using Multiagent Systems

A multiagent system (MAS) comprises two or more (intelligent) agents to follow a specific goal. The MAS is now a research reality, and MASs are rapidly having a critical presence in many areas and environments. Interested readers can find some detailed reviews in several references.[1-5] In the last two decades, MASs have been widely used in many fields of control engineering, such as power system control, manufacturing/industrial control, congestion control, distributed control, hybrid control, robotics and formation control, remote control, and traffic control.

With the embedded learning capabilities, agents that are autonomous, proactive, and reactive are well suited for modeling and control of various real-world complex systems, including the power industry. The MAS philosophy and its potential value in power system applications have been discussed.[6-8] Major applications are in the area of simulation, modeling, and design of trading platforms in restructured electricity markets.[6,9,10]

Recent published research works on MAS-based AGC design are briefly reviewed in Chapter 3. In this chapter, first an introduction of MASs is presented, then a multiagent reinforcement-learning-based AGC scheme is introduced, and finally, the proposed methodology is examined on some power system examples.

7.1 Multiagent System (MAS): An Introduction

Dealing with complex dynamic systems that can be described with the terms *uncertainty, nonlinearity, information structure constraints*, and *dimensionality*, it is very difficult to satisfy all requirements of an intelligent control system, such as adaptation and learning, autonomy and intelligence, as well as structures and hierarchies, by using fuzzy controllers, neural networks, and evolutionary optimization methods like genetic algorithm as single applications. There are no concepts to incorporate all these methods in one common framework that combines the advantages of the single methods.[11] One approach to design an intelligent control system to autonomously achieve a high level of control objectives could be the application of multiagent systems.

Multiagent systems perform a subfield of (distributed) artificial intelligence (AI). An MAS includes several agents and a mechanism for coordination of independent agents' actions. Various definitions for an agent are given in the computer science and AI literature.[12,13] An *agent* can be considered an intelligent entity that is operating in an *environment*, with a degree of autonomy, specific goal(s), and knowledge.

An agent can alter the environment by taking some actions, and can act autonomously in response to environmental changes. Autonomy means that the agent is able to fulfill its tasks without the direct intervention of a human, and the environment is everything (systems, hardware, and software) external to the agent. Of course, the agent is also a part of the environment, and can alter the environment by taking some actions. In a single-agent system, if there are other agents, they are also considered part of the environment.

Suitability for representing and control of interconnected/distributed systems, simplicity of mechanism, programming, and implementation, the capability of parallel processing/computation, scalability (handling numerous units), extensibility and flexibility (integrating of new parts and entities), maintainability (because of modularity due to using multiple components—agents), responsiveness (handling anomalies locally instead of propagating them to the whole system), robustness against failure, and reliability are some important reasons to use MASs in (specifically distributed) control system designs.

Figure 7.1 illustrates a view of typical MAS conceptually. Here, the agent is shown as a unit that sends and receives messages and interacts (via sensors and actuators) with its environment autonomously. The agents may also interact directly, as indicated in the figure by the arrows between the agents.

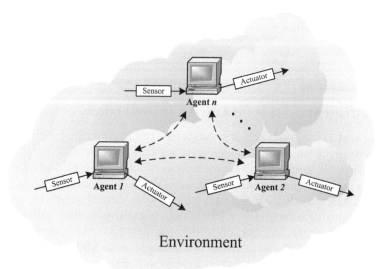

FIGURE 7.1
A general multiagent framework.

There may be numerous agents with different structures, local goals, actions, and domain knowledge—with or without the ability to communicate with other agents directly. In addition to autonomy, veracity, and rationality, the main characteristics an agent may have are social ability, responsiveness, proactiveness, adaptability, mobility, and learning. These characteristics are well defined in Wooldridge and Jennings.[2]

Over the years, various approaches to implement autonomous intelligent agents, such as belief-desire-intention (BDI) agents, reactive agents, agents with layered architectures,[13] and agents implemented using model-based programming,[14] have been introduced. The BDI approach is based on mental models of an agent's beliefs, desires, and intentions. It considers agents to have beliefs (about itself, other agents, and its environments), desires (about future states), and intentions (about its own future actions). Reactive agents are normally associated with the model of intelligence. The fundamental property of reactive agents is that they do not perform reasoning through interaction with the environment. Instead, they react to inputs from their environment and messages from other agents.[13]

Several layered agent structures are discussed in Wooldridge and Weiss.[13] In a layered agent, each layer is developed for a specific task. For example, consider agents with three layers:[15] a layer for handling measurement (message), a layer for behavior analysis, and a layer for actions. In this case, the message handling layer is responsible for sending and receiving messages from other agents, the behavioral layer will instruct the message handling layer to inform other agents of the new data, and the actions layer has the core functional attributes of the agent to perform the actions.[13]

The task of an agent contains a set of essential activities. Its goal is to change (or maintain) the state of the domain in some desirable way (according to the interest of its human principal). To do so, it takes action from time to time. To take the proper action, it makes observations on the domain. Following observations on a domain, an agent performs inference based on its knowledge about the relations among domain events, and then estimates the state of the domain using an intelligent core. The activity of guessing the state of the domain from prior knowledge and observations is known as *reasoning* or *inference*. A multiagent system consists of a set of agents acting in a problem domain. Each agent carries only a partial knowledge representation about the domain and can observe the domain from a partial perspective. Although an agent in a multiagent system can reason and act autonomously, as in the single-agent paradigm, to overcome its limit in domain knowledge, perspective, and computational resources, it can benefit from other agents' knowledge, perspectives, and computational resources through communication and coordination.[16]

In a multiagent system, at least one agent is usually equipped with intelligent inference. An intelligent core can play a major role in an intelligent agent for reasoning about the dynamic environment. Various intelligent cores/inferences, such as symbolic representation, if-then rules and fuzzy logic,[17] artificial neural networks,[18] reinforcement learning,[19] and Bayesian networks,[20] can be used in MASs. It may also possible to use some well-known control theory, such as sliding mode control, to suppress the effects of modeling uncertainties and disturbances, and to force the agent dynamics to move along a stabilizing manifold called *sliding manifold*.[21,22]

Multiagent control systems represent control schemes that are inherently distributed and consist of multiple entities/agents. The control architecture for MASs can be broadly characterized as *deliberative control, reactive control*, and a combination of both. Deliberative control is based on planning, while reactive control is based on coupling between sensing and actuation. Strategies that require that action be mediated by some symbolic representation of the environment are often called deliberative. In contrast, reactive strategies do not exhibit a steadfast reliance on internal models. Instead of responding to entities within a model, the control system can respond directly to the perception of the real world.[23]

Complex control tasks can always be hierarchically decomposed into several simpler subtasks. This naturally leads to agent architectures consisting of multiple layers. Figure 7.2 shows a schematic diagram of a three-layer agent architecture. In real-time control applications, the agents should

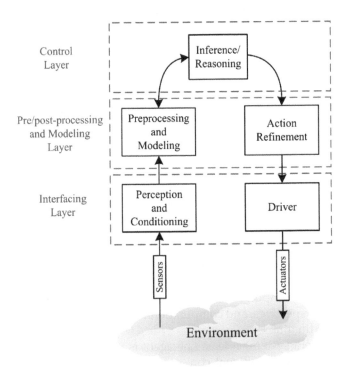

FIGURE 7.2
A typical intelligent agent architecture.

be capable of reasoning about the best possible action without losing too much time on sending or receiving data. In order to deal with the timing constraints caused by the real-time nature of the domain, it was therefore desirable that the agents could perform this high-level reasoning process. Adopting a somewhat hybrid approach thus seemed to be an appropriate choice.[24]

Figure 7.2 shows the functional hierarchy of the agent architecture. As illustrated in this figure, the agent and environment form a closed-loop system. The bottom layer is the *interfacing layer,* which takes care of the interaction with the environment. This layer observes the details of the environment as much as possible from the other layers. The middle layer is the signal *processing/modeling layer,* which simulates and provides a clear view of the world (environment) with a set of possible choices for a third layer. The highest layer in the architecture is the *control layer,* which contains the reasoning component of the system. In this layer, the best possible action is selected from the observation/modeling layer, depending on the current environment state and the current strategy of the overall control system. The most recent environment state information is then used by the control layer to reason about the best possible action. The action selected by the control layer is subsequently worked out in the second layer, which determines the appropriate actuator command. This command is then executed by the actuator control module in the interfacing layer.

7.2 Multiagent Reinforcement-Learning-Based AGC

One of the adaptive and nonlinear intelligent control techniques that can be effectively applicable in the power system AGC design is reinforcement learning (RL). Some efforts have been addressed.[25–29] The RL-based controllers learn and are adjusted to keep the area control error small enough in each sampling time of an AGC cycle. Since these controllers are based on learning methods, they are independent of environment conditions and can learn a wide range of operating conditions. The RL-based AGC design is model-free and can be easily scalable for large-scale systems and suitable in response to the load disturbances and power fluctuations.

The present section addresses the AGC design using an agent-based RL technique for an interconnected power system. Here, each control area includes an agent that communicates with others to control the frequency and tie-line power among the whole interconnected system. Each agent (control agent) provides an appropriate control action according to the area control error (ACE) signal, using an RL algorithm. In a multiarea power system, the learning process is considered a multiagent RL process, and agents of all areas learn together (not individually).

7.2.1 Multiagent Reinforcement Learning

This section presents a brief background on multiagent RL. The basic concepts and a comprehensive survey have been given previously.[19,30] The RL is learning what to do, and how to map situations to actions, so as to maximize a numerical reward signal.[19] In fact, the learner discovers which action should be taken by interacting with the environment and trying the different actions that may lead to the highest reward. The RL evaluates the actions taken and gives the learner feedback on how good the action taken was and whether it should repeat this action in the same situation. In other words, the RL methods as intelligent agents learn to solve a problem by interacting with their environments.

During the learning process, the agent interacts with the environment and takes an action a_t from a set of actions, at time t. These actions will affect the system and will take it to a new state x_{t+1}. Therefore, the agent is provided with the corresponding reward signal (r_{t+1}). This agent-environment interaction is repeated until the desired objective is achieved. A state signal indicates required information for making a decision and, if it succeeds in retaining all relevant information, is said to be Markov, or to have the Markov property,[19] and a RL task that satisfies this property is called a finite Markov decision process (MDP). If an environment has the Markov property, then its dynamics enable one to predict the next state and expected next reward given the current state and action.

In each MDP, the objective is to maximize the sum of returned rewards over time. Then the expected sum of discounted rewards is defined by the following equation:

$$R = \sum_{k=0}^{\infty} r^{k_{r+1}}$$

(7.1)

where γ is a discount factor that gives the most importance to the recent rewards, and $0 < \gamma < 1$.

Another term is a value function that is defined as the expected return or reward (E) when starting at state x_t while following policy $\pi(x,a)$ (see Equation 7.2). The policy shows the way the agent maps the states to the actions.[36]

$$V^{\pi}(x) = E_{\pi}\left\{ \sum_{k=0}^{\infty} r^{k_{t+k+1}} \mid x_t = x \right\}$$

(7.2)

An optimal policy is one that maximizes the value function. Therefore, once the optimal state value is derived, the optimal policy can be found as follows:

$$V^*(x) = \max_{\pi} V^{\pi}(x), \forall x \in X$$

(7.3)

In most RL methods, instead of calculating the state value, another term, known as the action value, is calculated (Equation 7.4), which is defined as the expected discounted reward while starting at state x_t and taking action a_t.

$$Q^\pi(x,a) = E_\pi \left\{ \sum_{k=0}^{\infty} \gamma^k r_{t+k+1} \Big| x_t = x, a_t = a \right\} \qquad (7.4)$$

To calculate the optimal action value, Bellman's equation,[19] as shown in Equation 7.5, can be used. In general, an optimal policy is one that maximizes the Q-function, defined by

$$Q^*(x,a) = \max_\pi E_\pi \left\{ r_{t+1} + \gamma \max_{\hat{a}} Q^*(x_{t+1}, \hat{a}) \Big| x_t = x, a_t = a, \right\} \qquad (7.5)$$

Different RL methods have been proposed to solve the above equations. In some algorithms, the agent will first approximate the model of the system in order to calculate the Q-function. The method used in the present chapter is of a temporal difference type, which learns the model of the system under control. The only available information is the reward achieved by each action taken and the next state. The algorithm, called Q-learning, will approximate the Q-function, and by the computed function the optimal policy, which maximizes this function, is derived.[27]

Well-understood algorithms with desirable convergence and consistency properties are available for solving the single-agent RL task, both when the agent knows the dynamics of the environment and the reward function, and when it does not. However, the scalability of algorithms to realistic problem sizes is problematic in single-agent RL, and is one of the great reasons to use multiagent RL.[30] In addition to scalability and benefits owing to the distributed nature of the multiagent solution, such as parallel processing, multiple RL agents may utilize new benefits from sharing experience, e.g., by communication, teaching, or imitation.[30] These properties make RL attractive for multiagent learning.

However, several new challenges arise for RL in the MASs. In MASs, other adapting agents make the environment no longer stationary, violating the Markov property that traditional single-agent behavior learning relies on; these nonstationarity properties decrease the convergence properties of most single-agent RL algorithms.[31] Another problem is the difficulty of defining an appropriate learning goal for the multiple RL agents.[30] Only then, an RL agent will be able to coordinate its behavior with other agents. These challenges make the multiagent RL design and learning difficult in large-scale applications; one should use a special learning algorithm, such as those introduced in Bevrani et al.[28] and Busoniu et al.[30] and discussed in Section 7.2.3. Using such a learning algorithm, violation of the Markov property caused from a multiagent structure and other problems will be solved.

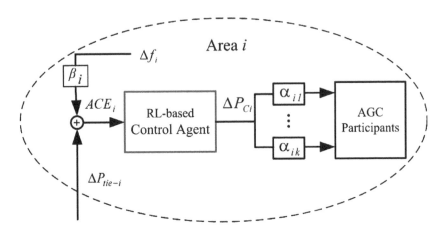

FIGURE 7.3
The overall control framework for area i.

7.2.2 Area Control Agent

Figure 7.3 shows the proposed intelligent control framework for control area i in a multiarea power system. Each control area includes an RL-based control agent as an intelligent controller. The controller is responsible for producing an appropriate control action (ΔP_{Ci}) using RL, according to the measured ACE signal and tie-line power changes (ΔP_{tie-i}).

The intelligent controller (control agent) functions as follows: At each instant (on a discrete timescale k, $k = 1, 2, ...$), the control agent observes the current state of the system x_k and takes an action a_k. The state vector consists of some quantities, which are normally available to the control agent. Here, the average of the ACE signal over the time interval $k - 1$ to k as the state vector at the instant k is used. For the algorithm presented here, it is assumed that the set of all possible states X is finite. Therefore, the values of various quantities that constitute the state information should be quantized.

The possible actions of the control agent are the various values of ΔP_C that can be demanded in the generation level within an AGC interval. The ΔP_C is also discretized to some finite number of levels. Now, since both X and $A = \{a_k; k = 1, 2, ...\}$ are finite sets, a model for this dynamic system can be specified through a set of probabilities.

7.2.3 RL Algorithm

Here, similar to the introduced algorithm in Ahamed et al.,[25] an RL algorithm is used for estimating Q^* and the optimal policy (Equation 7.5). Suppose we have a sequence of samples (x_k, x_{k+1}, a_k, r), $k = 1, 2, \cdots$. Each sample is such that x_{k+1} is a random state that results when action a_k is performed in state x_k, and $r_k = g(x_k, x_{k+1}, a_k)$ is the consequent immediate reinforcement.

Such a sequence of samples can be obtained through either a simulation model of the system or observing the actual system in operation. This sequence of samples (called training set) can be used to estimate Q^*, using a specific algorithm. Suppose Q^k is the estimated value of Q^* at the kth iteration. Let the next sample be (x_k, x_{k+1}, a_k, r), then Q^{k+1} can be obtained as follows:

$$Q^{k+1}(x_k, a_k) = Q^k(x_k, a_k) + \alpha \left[g(x_k, x_{k+1}, a_k) + \gamma \max_{\alpha \in A} Q^k(x_{k+1}, \hat{a}) - Q^k(x_k, a_k) \right] \quad (7.6)$$

where a is a constant called the step size of learning algorithm, and $0 < a < 1$.

At each time step, as determined by the sampling time for the AGC action, the state input vector x for the AGC is determined, and then an action in that state is selected and applied to the model. The model is integrated for a time interval equal to the sampling time of AGC to obtain the state vector $\hat{x}$ at the next time step.

Here, the exploration policy for choosing actions in different states is used. It is based on a learning automata algorithm called *pursuit algorithm*.[32] This is a stochastic policy where, for each state x, actions are chosen based on a probability distribution over the action space. Let P_x^k denote the probability distribution over the action set for state vector x at the kth iteration of learning; that is, $P_x^k(a)$ is the probability of choosing action a in state x at iteration k. An uniform probability distribution is considered at $k = 0$, that is,

$$P_x^0(a) = \frac{1}{|A|} \forall \alpha \in A \;\; \forall x \in X \quad (7.7)$$

At the kth iteration, let the state x_k be equal to x. An action a_k based on $P_x^k(.)$ is randomly chosen. That is, $Prob(a_k = a) = P_x^k(a)$. Using the performed simulation model, the system goes to the next state, x_{k+1}, by applying action a in state x, and is integrated for the next time interval. Then, Q^k is updated to Q^{k+1} using Equation 7.6, and the probabilities are updated as follows:

$$P_x^{k+1}(a_g) = P_x^k(a_g) + \beta(1 - P_x^k(a_g))$$

$$P_x^{k+1}(a) = P_x^k(a)(1 - \beta) \;\; \forall \alpha \in A, a \neq a_g \quad (7.8)$$

$$P_y^{k+1}(a) = P_y^k(a) \;\; \forall \alpha \in A, \forall y \in X, y \neq x$$

where β is a constant and $0 < \beta < 1$. Thus, at iteration k, the probability of choosing the greedy action a_g in state x is slightly increased, and the probabilities of choosing all other actions in state x are proportionally decreased.

In the present algorithm, the aim is to achieve the well-known AGC objective and to keep the ACE within a small band around zero. This choice is motivated by the fact that all the existing AGC implementations use this as a main control objective, and hence it will be possible to compare the proposed RL approach with conventional and other AGC design approaches.

As mentioned earlier, in this formulation, each state vector consists of the average value of ACE as a state variable. The control agent actions change the generation set point, ΔP_C. According to the RL algorithms application, usually a finite number of states are assumed. In this direction, the state variable and action variable should be discretized to finite levels, too.

The next step is to choose an immediate reinforcement function by defining the function g. The reward matrix initially is full of zeros. At each time step, the average value of the ACE signal is obtained; then, according to its discretized values, the state of the system is determined. Whenever the state is desirable (i.e., $|ACE|$ is less than ε), the reward function $g(k, x_{k+1}, a_k)$ is assigned at zero value. When it is undesirable (i.e., $|ACE| > \varepsilon$), $g(k, x_{k+1}, a_k)$ is assigned a value $-|ACE|$. In this process, all actions that cause an undesirable state with a negative value are penalized.

7.2.4 Application to a Thirty-Nine-Bus Test System

To illustrate the effectiveness of the proposed control strategy, the designed intelligent control scheme is applied to the thirty-nine-bus test system described in Figure 6.7. The power system is divided into three control areas, as explained in Chapter 6. Here, the purpose is essentially to clearly show the various steps of implementation and illustrate the method. After design choices are made, the controller is trained by running the simulation in the learning mode, as explained in the previous section. After completing the learning phase, the control actions at various states converge to their optimal values.

The simulation is run as follows: At each AGC instant k, the control agents of all areas average all corresponding ACE signal instances gained every 0.1 s. Three average values of ACE signal instances, each related to one area, form the current state vector, x_k, that is obtained according to the quantized states. When all areas' state vectors are ready, the control agents choose the action signal a_k that consists of three ΔP_C values for three areas (action signal is gained according to the quantized actions and the exploration policy mentioned above) to change the set points of the governors using the values given by a_k.

In the performed simulation studies, the input variable is obtained as follows. As the AGC decision time cycle chosen, three values of ACE (for three control areas) are calculated over the determined cycle. The averages of these values for three areas are the state variable ($x^1_{avg1}, x^1_{avg2}, x^1_{avg3}$).

Since in the multiagent RL process the agents of all areas are learning together, the state vector also consists of all state vectors of three areas, and the action vector consists of all action vectors of three areas, as shown in $<(X_1, X_2, X_3), (A_1, A_2, A_3), (r_1, r_2, r_3)>$ or $<X, A, p, r>$. Here $X_i = x^1_{avgi}$ is the discrete set of each area state, X is the joint state, A_i is the discrete set of each area action available to the area i, and A is the joint action. In each time instant after averaging of ACE_i for each area (over three instances), depending on the current joint state (X_1, X_2, X_3), the joint action ($\Delta P_{C1}, \Delta P_{C2}, \Delta P_{C3}$) is chosen according to the exploration policy.

Consequently, the reward r also depends on the joint action whenever the next state (X) is desirable (i.e., all $|ACE_i|$ are less than ε); then reward function r is fixed at zero value. When the next state is undesirable (i.e., $\exists\, ACE_i,\ |ACE| > \varepsilon$), r is assigned an average value of $-|ACE_i|$. In this algorithm, since all agents learn together, parallel computation causes the learning process to speed up. This RL algorithm is also more scalable than single-agent RL algorithms.

In the performed simulations, the proposed controllers are applied to the thirty-nine-bus, three-control area system, as simplified in Figure 7.4. In this section, the performance of the closed-loop system using the well-tuned conventional PI controllers is compared to that of the system using the designed multiagent RL controllers for a wide range of load disturbances.

As a serious test scenario, similar to in Section 6.3.2, the following load disturbances (step increase in demand) are applied to three areas: 3.8% of the total area load at bus 8 in area 1, 4.3% of the total area load at bus 3 in area 2, and 6.4% of the total area load at bus 16 in area 3 have been simultaneously increased in step form. The applied step load disturbances ΔP_{Li} (pu), the output power of wind farms P_{WT} (MW), and the wind velocity V_W (m/s) are shown in Figure 7.5.

The frequency deviation (Δf) and area control error (ACE) signals in three areas are shown in Figures 7.6 and 7.7, respectively. The produced mechanical power by the AGC participating unit in area 2 (P_{m2} for G_9), the corresponding electrical power (P_{e2}), and also the overall tie-line power for the same area (P_{tie2}) are shown in Figure 7.8.

The wind penetration in this system is considered as two individual wind farms, each with a capacity equivalent to about half of the total penetration.

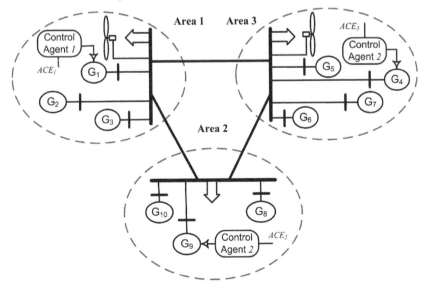

FIGURE 7.4
Three-control area with RL-based control agents.

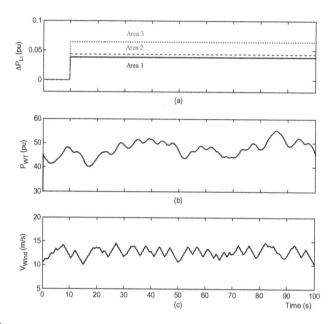

FIGURE 7.5

(a) Load step disturbances in three areas, (b) total wind power, and (c) the wind velocity pattern in area 1.

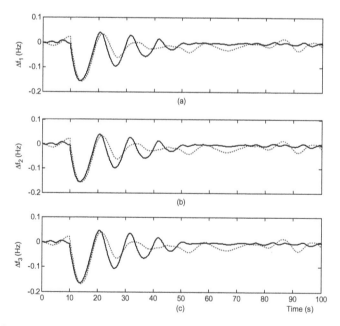

FIGURE 7.6

Frequency deviation in (a) Area 1, (b) Area 2, and (c) Area 3. Proposed intelligent method (solid), linear PI control (dotted).

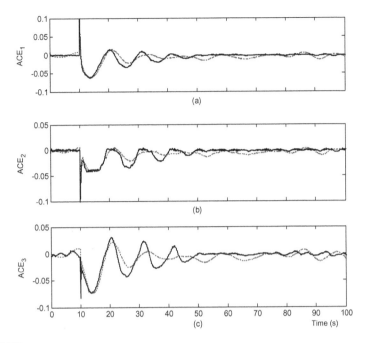

FIGURE 7.7
ACE signal in (a) Area 1, (b) Area 2, and (c) Area 3. Proposed intelligent method (solid), linear PI control (dotted).

However, in the present simulation, the detailed dynamic nonlinear models of a thirty-nine-bus power system and wind turbines are used without applying an aggregation model for generators or wind turbine units. That is why, in the simulation results, in addition to the long-term fluctuations, some fast oscillations in a timescale of 10 s are also observable.[28,33]

As shown in the simulation results, using the proposed method, the area control error and frequency deviation of all areas are properly driven close to zero. Furthermore, regarding that the proposed algorithm is an adaptive algorithm and is based on the learning methods, i.e., in each state it finds the local optimum solution to gain the system objective (minimizing the ACE signal), the intelligent controllers provide smoother control action signals, and area frequency deviations are less than the frequency deviations in the same system with conventional controllers.

7.3 Using GA to Determine Actions and States

The genetic algorithm (GA) can be used to gain better results and to tune the quantized values of the state vector and action vectors. To quantize the state range and action range using GA, each individual that is an explanatory

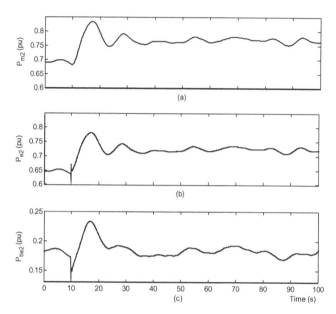

FIGURE 7.8
Area 2 power response using the proposed multiagent RL method.

quantized value of states and actions should be a double vector. It is clear that with increasing the number of variables in a double vector, the states (ACE signal quantized values) are found more precisely. In the AGC issue, system states are more important than system actions (ΔP_C quantized values) and have a greater affect on the whole system performance (keeping the ACE within a small band around zero), because the systems with more states can learn more precisely than the same systems with less states.

First, the maximum number of states (n_s) and the least valuable actions number (n_a) should be defined in GA for the assumed AGC system. For the actions variable, in case of considering a small number of variables, the learning speed will increase, because it is not necessary to examine extra actions in each state.[26]

7.3.1 Finding Individual's Fitness and Variation Ranges

To find eligibility (fitness) of individuals, n_a variables are randomly chosen as discretized values of actions from each individual, which contains ($n_a + n_s$) variables; then these values should be scaled according to the action range (variable's range is between 0 and 1; however, the variation of the ΔP_C action signal is between [ΔP_{Cmin} ΔP_{Cmax}]), and the remaining other n_s variables are discretized values of the ACE signal, which should be scaled to the valid range ([ACE_{min} ACE_{max}]). After scaling and finding the corresponding quantized state and action vector, the model is run with these properties, and the individual's fitness is obtained from Equation 7.9. The individuals with the smallest fitness are the best.

$$Individual\ Fitness = \sum |ACE| / (simulation\ time) \qquad (7.9)$$

Hence, the basic AGC task can be summarized to correct the observed ACE within a limited range; if the ACE goes beyond this range, other emergency control steps may have to be taken by the operator. Let the valuable range of ACE for which the AGC is expected to act properly be $[ACE_{min}\ ACE_{max}]$. In fact, ACE_{min} and ACE_{max} must be determined by the operating policy of the area according to the existing frequency operating standards.[33] ACE_{max} is the maximum ACE signal deviation that is expected to be corrected by the AGC (in practice, ACE deviations beyond this value are corrected only through operator intervention or may need an emergency control action). ACE_{min} is the required minimum amplitude of *ACE* deviation to trigger the AGC control loop.

The other variable to be quantized is the control action ΔP_C. This also requires that a design choice be made for the range $[\Delta P_{Cmin}\ \Delta P_{Cmax}]$. The ΔP_{Cmax} is automatically determined by the equipment constraints of the system. It is the maximum power change that can be effected within one AGC execution period. The ΔP_{Cmin} is the minimum change that must be demanded according the dynamics of the AGC participating units. The following application example illustrates how the GA can be practically used to determine actions and states in a multiagent RL-based AGC design. Interested readers can find more details in Daneshfar and Bevrani[26] and Daneshfar.[34]

7.3.2 Application to a Three-Control Area Power System

To illustrate the effectiveness of using GA in the RL algorithm for the proposed control strategy, a three-control area power system (same as the example used in Section 2.4 and Bevrani et al.[35]) is considered as a test system. Each control area includes three Gencos, and the power system parameters are given in Table 2.1. The schematic diagram of the system used for simulation studies is also shown in Figure 2.13.

After completing the design steps of the algorithm, the controller must be trained by running the simulation in the learning mode, as explained in Section 7.2. The performance results presented here correspond to the performance of the controllers after ending the learning phase and converging the controller's actions at various states to their optimal values. As described in a previous example (Section 7.2.4), at each AGC execution period that is greater than the simulation sampling time, the control agent of each area averages all corresponding ACE signal instances measured by sensors and averages all load change instances obtained during the AGC execution period. Three average values of ACE signal instances for three areas, together with three average values of load change instances, form the current joint state vector x_k and are obtained according to the quantized states gained from GA. Then, the control agents choose an action a_k according to the quantized actions gained from GA and the aforementioned exploration policy. Each joint action a_k consists of three actions (ΔP_{C1}, ΔP_{C2}, ΔP_{C3}) to change the set points of the governors. Using these actions for the governors setting, the AGC process is transferred to the next execution period. During the next cycle (i.e., until the next instant of AGC gained), three values of average ACE instances in each area are formed for the next joint state x_{k+1}.

In the presented simulation study, the input variable is obtained as follows: At each AGC execution period, average values of ACE signal instances corresponding to three areas are calculated; they are the first state variables (x^1_{avg1}, x^1_{avg2}, x^1_{avg3}). In the multiagent RL process, agents of all areas are learning together, the joint state vector consists of all state vectors of three areas, and the joint action vector consists of all action vectors of three areas. This statement can be shown in triple $<(X_1, X_2, X_3), (A_1, A_2, A_3), p, (r_1, r_2, r_3)>$ or $<X, A, p, r>$, where $X_i = (x^1_{avgi}, x^2_i)$ is the discrete set of each area state, X is the joint state, A_i is the discrete sets of actions available to area i, and A is the joint action.

In each AGC execution period, after averaging of ACE_i of all areas (over instances obtained in that period), depending on the current joint state (X_1, X_2, X_3), the joint action (ΔP_{C1}, ΔP_{C2}, ΔP_{C3}) is chosen according to the exploration policy. Consequently, the reward r also depends on the joint action whenever the next state (X_i) is desirable, i.e., all $|ACE_i|$ are less than ε, where ε is the smallest ACE signal value that AGC can operate. Then, the reward function r is assigned a zero value. When the next state is undesirable, i.e., at least one $|ACE_i|$ is greater than ε, r is assigned an average value of all $-|ACE_i|$.

For the sake of simulations, the performance of the closed-loop system for the mentioned three-control area using the linear robust proportional-integral (PI) controllers,[35] compared to the designed multiagent RL controllers, is tested for the following simultaneous large load disturbances (step increase in demand) in three areas:

$$\Delta P_{L1} = 100MW; \Delta P_{L2} = 80MW; \Delta P_{L3} = 50MW$$

The frequency deviation, ACE signal, and control action signals of the closed-loop system are shown in Figures 7.9 to 7.11. Simulation results show that the ACE and frequency deviation of all areas for the proposed intelligent GA-based multiagent RL controllers are properly driven back to zero, as well as for robust PI controllers. The produced control action signals, which are proportional to the specified participation factors, are smooth enough to satisfy the generation physical constraint.

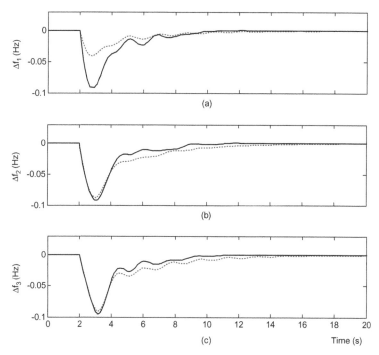

FIGURE 7.9
Frequency deviation in (a) Area 1, (b) Area 2, and (c) Area 3. Proposed intelligent method (solid), robust PI control (dotted).

7.4 An Agent for β Estimation

The frequency bias factor (β) is an important term to calculate ACE (Equation 2.11). Since the control agents in the described AGC scheme provide control action signals based on the received ACE signals to achieve more accurate results, one may use an individual agent in each area for estimation of β. The conventional approaches in tie-line bias control use the frequency bias coefficient $-10B$ to offset the area's frequency response characteristic, β. But it is related to many factors, and with $-10B = β$, the ACE would only react to internal disturbances. Therefore, recently several approaches have been given to approximate β instead of a constant value for real-time applications.[36–39]

A multiagent AGC scheme including an estimator agent to estimate the β parameter is given in Daneshfar and Bevrani.[26] The overall control framework is shown in Figure 7.12. The estimator agent in each control area calculates the β parameter and performs ACE based on received signals ΔP_{tie}, Δf, ΔP_m, ΔP_L. The estimation algorithm is developed based on a dynamical

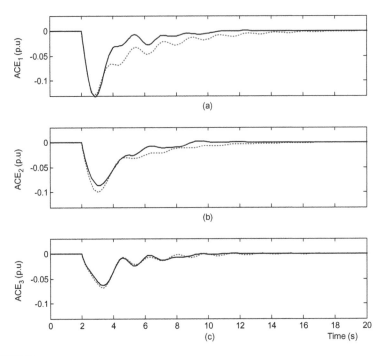

FIGURE 7.10
ACE signal in (a) Area 1, (b) Area 2, and (c) Area 3. Proposed intelligent method (solid), robust
PI control (dotted).

representation of generation-load in the simplified AGC frequency response
model (Figure 2.8), which can be described as follows:

$$\sum_{j=1}^{n} \Delta P_{mji}(t) - \Delta P_{Li}(t) - \Delta P_{tie,i}(t) = 2H_i \frac{d}{dt} \Delta f_i(t) + D_i \Delta f_i(t) \qquad (7.10)$$

Substituting $\Delta P_{tie,i}(t)$ from Equation 2.1 in Equation 7.10 yields

$$\sum_{j=1}^{n} \Delta P_{mji}(t) - \Delta P_{Li}(t) + \beta_i \Delta f_i(t) - ACE_i(t) = 2H_i \frac{d}{dt} \Delta f_i(t) + D_i \Delta f_i(t) \qquad (7.11)$$

From Equation 7.11, ACE can be calculated in terms of other variables:

$$ACE_i(t) = \sum_{j=1}^{n} \Delta P_{mji}(t) - \Delta P_{Li}(t) + (\beta_i - D_i)\Delta f_i(t) - 2H_i \frac{d}{dt} \Delta f_i(t) \qquad (7.12)$$

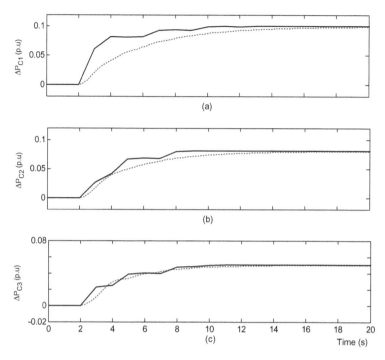

FIGURE 7.11
Control action signal in (a) Area 1, (b) Area 2, and (c) Area 3. Proposed intelligent method (solid), robust PI control (dotted).

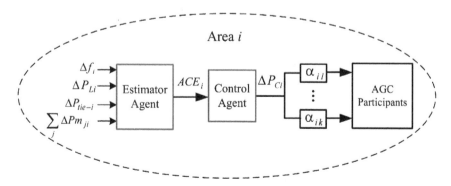

FIGURE 7.12
Control framework for area i, with estimator agent.

By applying the following definition for a moving average over a T second interval to Equation 7.12,

$$\bar{X}_T = \frac{1}{T}\int_{t_i}^{t_f} X(t)dt, \quad t_f - t_i = T \; second \tag{7.13}$$

$\overline{ACE}_T$ can be obtained as follows:

$$\overline{ACE}_T = \frac{1}{T}\left\{\sum_T\sum_{j=1}^{n}\Delta P_{mji} - \sum_T \Delta P_{Li} + (\beta_i - D_i)\sum_T \Delta f_i\right\} - \frac{2H_i}{T}(\Delta f(t_f) - \Delta f(t_i)) \tag{7.14}$$

By applying the measured values of ACE and other variables in the above equation over a time interval, the values of β can be estimated for the corresponding period. Since the values of β vary with system conditions, these model parameters would have to be updated regularly using a recursive least squares (RLS) algorithm.[40] Suitable values of the duration T depend on system dynamic behavior. Although a larger T would yield smoother β values, the convergence to the proper value would be slow. Figure 7.13 shows the estimated and calculated β over a 100 s simulation, for area 1 of

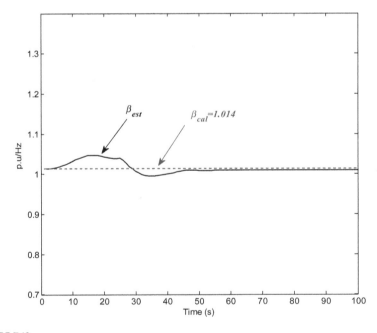

FIGURE 7.13
The response of estimator agent.

the three-control area described in Section 7.3.2. For this test the $-10B$ of the target control area was set equal to β_{cal}. As it is clear, β_{est} converged rapidly to the β_{cal} and remained there over the rest of the run. Several simulation scenarios for the example at hand using an estimator agent are presented in Daneshfar and Bevrani.[26] The application results of using GA to optimize actions and states in the multiagent RL approach for the thirty-nine-bus test system are given in Daneshfar and Bevrani[26] and Daneshfar.[34]

7.5 Summary

This chapter addresses the application of MASs in AGC design for interconnected power systems. General frameworks for agent-based control systems based upon the foundations of agent theory are discussed. A new multiagent AGC scheme has been introduced. The capability of reinforcement learning in the proposed AGC strategy is examined, and the application of GA to determine actions and states during the learning process is discussed. The possibility for building of more agents such as estimator agents to cope with real-world AGC systems is explained.

Model independency, scalability, flexibility, the decentralized property, and the capability of parallel processing, as main features of new approaches in specifying the control objective, make it very attractive for application in AGC design. The application results in some power system examples show that the RL-based multiagent control schemes perform well, in comparison to the performance of robust PI control design. In Chapter 8, another multiagent AGC scheme based on Bayesian networks is presented.

References

1. K. P. Sycara. 1998. Multiagent systems. *AI Magazine* 19(2):79–92.
2. M. Wooldridge, N. Jennings. 1995. Intelligent agents: Theory and practice. *Knowledge Eng. Rev.* 10(2):115–52.
3. G. Weiss. 1999. *Multiagent systems: A modern approach to distributed artificial intelligence.* Cambridge, MA: MIT Press.
4. P. Stone, M. Veloso. 2000. Multiagent systems: A survey from a machine learning perspective. *Autonomous Robots* 8(3):345–83.
5. M. Wooldridge, N. R. Jennings. 1995. Agent theories, architectures, and languages: A survey. In *Intelligent agents*, 1–39. Berlin: Springer.
6. Power Systems Engineering Research Center (PSERC). 2008. *Agent modeling for integrated power systems.* Final project report. PSERC. http://www.pserc.wisc.edu/documents/publications/.

7. S. D. J. McArthur, E. M. Davidson, V. M. Catterson, A. L. Dimeas, N. D. Hatziargyriou, F. Ponci, T. Funabashi. 2007. Multi-agent systems for power engineering applications. Part I. Concepts, applications and technical challenges. *IEEE Trans. Power Syst.* 22(4):1743–52.
8. S. D. J. McArthur, E. M. Davidson, V. M. Catterson, A. L. Dimeas, N. D. Hatziargyriou, F. Ponci, T. Funabashi. 2007. Multi-agent systems for power engineering applications. Part II. Technologies, standards and tools for multi-agent systems. *IEEE Trans. Power Syst.* 22(4):1753–59.
9. P. Wei, Y. Yan, Y. Ni, J. Yen, F. Wu. 2001. A decentralized approach for optimal wholesale cross-border trade planning using multi-agent technology. *IEEE Trans. Power Syst.* 16:833–38.
10. A. L. Dimeas, N. D. Hatziargyriou. 2005. Operation of a multiagent system for microgrid control. *IEEE Trans. Power Syst.* 20:1447–55.
11. H. Voos. 2000. Intelligent agents for supervision and control: A perspective. In *Proceedings of the 15th IEEE International Symposium on Intelligent Control (ISIC)*, Rio Patras, Greece, pp. 339–44.
12. S. Russell, P. Norvig. 1995., *Artificial intelligence: A modern approach.* Englewood Cliffs, NJ: Prentice-Hall.
13. M. Wooldridge, G. Weiss, eds. 1999. *Intelligent agents, in multi-agent systems,* 3–51. Cambridge, MA: MIT Press.
14. B. C. Williams, M. D. Ingham, S. H. Chung, P. H. Elliott. 2003. Model-based programming of intelligent embedded systems and robotic space explorers. *Proc. IEEE* 91(1):212–37.
15. F. Bellifemine, A. Poggi, G. Rimassa. 2001. Developing multi-agent systems with jade. In *Intelligent agents,* ed. C. Castelfranchi, Y. Lesperance, 89–103. Vol. VII, no. 1571 of Lecture Notes in Artificial Intelligence. Heidelberg, DE: Springer-Verlag.
16. Y. Xiang. 2002. *Probablistic reasoning in multiagent systems: A graphical models approach.* Cambridge: Cambridge University Press.
17. D. Poole, A. Mackworth, R. Goebel. 1998. *Computational intelligence: A logical approach.* New York: Oxford University Press.
18. M. H. Hassoun. 1995. *Fundamentials of artificial neural networks.* Cambridge, MA: MIT Press.
19. R. S. Sutton, A. G. Barto. 1998. *Reinforcement learning: An introduction.* Cambridge, MA: MIT Press.
20. H. Bevrani, F. Daneshfar, P. R. Daneshmand, T. Hiyama. 2010. Intelligent automatic generation control: Multi-agent Bayesian networks approach. In *Proceedings of IEEE International Conference on Control Applications,* Yokohama, Japan, CD-ROM.
21. V. Gazi, B. Fidan. 2007. Coordination and control of multi-agent dynamic systems: Models and approaches. In *Swarm robotics,* ed. E. Sahin, et al., 71–102. LNCS 4433. Berlin Heidelberg: Springer-Verlag.
22. V. I. Utkin. 1992. *Sliding modes in control and optimization.* Berlin: Springer-Verlag.
23. H. Li. 2006. A framework for coordinated control of multi-agent systems. PhD thesis, University of Waterloo, Waterloo, Ontario, Canada.
24. R. de Boer, J. Kok. 2002. The incremental development of a synthetic multi-agent system: The UvA trilearn 2001 robotic soccer simulation team. Master's thesis, Faculty of Science, University of Amsterdam.
25. T. P. I Ahamed, P. S. N. Rao, P. S. Sastry. 2006. Reinforcement learning controllers for automatic generation control in power systems having reheat units with GRC and dead-band. *Int. J. Power Energy Syst.* 26:137–46.

26. F. Daneshfar, H. Bevrani. 2010. Load-frequency control: A GA-based multi-agent reinforcement learning. *IET Gener. Transm. Distrib.* 4(1):13–26.

27. S. Eftekharnejad, A. Feliachi. 2007. Stability enhancement through reinforcement learning: Load frequency control case study. *Bulk Power Syst. Dynamics Control* VII:1–8.

28. H. Bevrani, F. Daneshfar, P. R. Daneshmand. 2010. Intelligent power system frequency regulation concerning the integration of wind power units. In *Wind power systems: Applications of computational intelligence*, ed. L. F. Wang, C. Singh, A. Kusiak, 407–37. Springer Book Series on Green Energy and Technology. Heidelberg: Springer-Verlag.

29. H. Bevrani, F. Daneshfar, P. R. Daneshmand, T. Hiyama. 2010. Reinforcement learning based multi-agent LFC design concerning the integration of wind farms. In *Proceedings of IEEE International Conference on Control Applications*, Yokohama, Japan, CD-ROM.

30. L. Busoniu, R. Babuska, B. de Schutter. 2008. A comprehensive survey of multi-agent reinforcement learning. *IEEE Trans. Syst. Man. Cyber. C Appl. Rev.* 38:156–72.

31. E. F. Yage, D. B. Gu. 2004. *Multi-agent reinforcement learning for multi-robot systems: A survey*. Technical Report CSM-404, University of Essex, Colchester, UK.

32. M. A. L. Thathachar, B. R. Harita. 1998. An estimator algorithm for learning automata with changing number of actions. *Int. J. Gen. Syst.* 14:169–84.

33. H. Bevrani. 2009. *Robust power system frequency control*. New York: Springer.

34. F. Daneshfar. 2009. Automatic generation control using multi-agent systems. MSc dissertation, Department of Electrical and Computer Engineering, University of Kurdistan, Sanandaj, Iran.

35. H. Bevrani, Y. Mitani, K. Tsuji. 2004. Robust decentralised load-frequency control using an iterative linear matrix inequalities algorithm. *IEE Proc. Gener. Transm. Distrib.* 3(151):347–54.

36. N. B. Hoonchareon, C. M. Ong, R. A. Kramer. 2002. Feasibility of decomposing ACE to identify the impact of selected loads on CPS1 and CPS2. *IEEE Trans. Power Syst.* 22(5):752–56.

37. L. R. Chang-Chien, N. B. Hoonchareon, C. M. Ong, R. A. Kramer. 2003. Estimation of β for adaptive frequency bias setting in load frequency control. *IEEE Trans. Power Syst.* 18(2):904–11.

38. L. R. Chang-Chien, C. M. Ong, R. A. Kramer., 2002. Field tests and refinements of an ace model. *IEEE Trans. Power Syst.* 18(2):898–903.

39. L. R. Chang-Chien, Y. J. Lin, C. C. Wu. 2007. An online approach to allocate operating reserve for an isolated power system. *IEEE Trans. Power Syst.* 22(3):1314–21.

40. E. K. P. Chong, S. H. Zak. 1996. *An introduction to optimization*. New York: John Wiley & Sons Press.

8

Bayesian-Network-Based AGC Approach

As discussed in Chapter 6, the AGC is becoming more significant today because of the increasing renewable energy sources, such as wind farms. The power fluctuation caused by a high penetration of wind farms negatively contributes to the power imbalance and frequency deviation. In this chapter, a new intelligent agent-based control scheme, using Bayesian networks (BNs), to design an AGC system in a multiarea power system is addressed. Model independency and flexibility in specifying the control objectives identify the proposed approach as an attractive solution for AGC design in a real-world power system. The BNs also provide a robust probabilistic method of reasoning under uncertainty. Efficient probabilistic inference algorithms in the BNs permit answering various probabilistic queries about the system. Moreover, using a multiagent structure in the proposed control framework realizes parallel computation and a high degree of scalability.

Currently, wind is the most widely utilized renewable energy technology in power systems. Wind turbine generators have attracted accelerated attention in recent years. Nowadays, due to the interconnection of more distributed generators, especially wind turbines, the electric power industry has become more complicated than ever. Since the primary energy source (wind) cannot be stored and is uncontrollable, the controllability and availability of wind power significantly differs from conventional power generation. In most power systems, the output power of wind turbine generators varies with wind speed fluctuation, and this fluctuation results in frequency variation.[1,2] Some reports have recently addressed the power system frequency control and AGC issues in the presence of wind turbines.[1,3–12] The AGC is known as one of the important future power system control problems concerning the integration of wind power turbines in multiarea power systems.

In response to the existing challenge of integrating computation, communication, and control into appropriate levels of the AGC system, this chapter introduces an intelligent BN-based multiagent control scheme to satisfy AGC objectives concerning the integration of wind power units.

The multiagent system concept and its great potential value to AGC systems are discussed in Chapter 7. On the other hand, BNs[13] provide a useful adaptive and nonlinear control technique that can be easily applicable in the AGC design. The BN is known as a powerful tool for knowledge representation and inference in control systems with uncertainties and undefined dynamics. It has been successfully applied in a variety of real-world engineering tasks, but has received little attention in the area of power system control issues.[14–16] It has been effectively used to incorporate expert knowledge and historical data for revising the prior belief in the light of new evidence in many fields. The main feature of the BN is the possibility of including local conditional dependencies into the model, by directly specifying the causes that influence a given effect.[15]

In this chapter, the proposed BN-based multiagent AGC framework includes two agents in each control area for estimating the amount of power imbalance and providing an appropriate control action signal according to load disturbances and tie-line power changes. The main advantages of the proposed BN scheme for the AGC application can be summarized as follows: (1) simplicity and intuitive model building that is closely based on the physical power system topology, (2) easy incorporation of uncertainty and dependency in the frequency response model, (3) capability to monitor the probability of any variable in the whole system, (4) propagation of probabilistic information that allows a large range of what-if analysis that is useful in wide area monitoring and control, and (5) independent of the power system parameter values, such as frequency bias factor.

To demonstrate the efficiency of the proposed control method, some nonlinear simulations on the New England ten-machine, thirty-nine-bus test system concerning the integration of wind power units are performed. A real-time laboratory experiment using the Analog Power System Simulator (APSS) at the Research Laboratory of the Kyushu Electric Power Company (Japan) is also performed. The results show that the proposed AGC scheme guarantees the optimal performance for a wide range of operating conditions.

The chapter is organized as follows: A brief introduction on BNs is given in Section 8.1. The AGC system with wind farms is discussed in Section 8.2. In Section 8.3, the proposed intelligent BN-based multiagent AGC scheme is presented. The BN's construction and parameter learning are explained in Section 8.4. Simulation results and laboratory experiments are provided in Section 8.5. Finally, the chapter is summarized in Section 8.6.

8.1 Bayesian Networks: An Overview

This section contains an overview on Bayesian networks (BNs), intended to provide readers with an understanding of these networks, including what they are and how they are used in the proposed intelligent AGC approach. For a technical review, interested readers are referred to more in-depth references.[13,17–19]

In real learning problems, there are a large number of variables with relationships. The BN is a suitable representation tool for such cases. A BN is a graphical model that efficiently encodes the joint probability distribution for a large set of variables. The BNs are widely used for representing uncertain knowledge in an artificial intelligence (AI) scheme. They have become the standard methodologies for the construction of systems relying on probabilistic knowledge and have been applied in a variety of real-world engineering problems.[20]

There are several attractive properties of BNs for the inference of power system fault diagnosis, reliability assessment, operation, and control. They can represent complex stochastic nonlinear relationships among multiple interacting dynamics, and their probabilistic nature can accommodate uncertainty inherent to measured/estimated data. They can describe direct dynamic interactions as well as indirect influences that proceed through additional, unobserved components, a property crucial for discovering previously unknown dynamics, effects, events, and unknown components. Therefore, very complex relationships that likely exist in a large-scale power system can be modeled and discovered. The BNs' inference algorithm constructs a graph diagram in which nodes represent the measured/estimated variables (states) and the lines between nodes (arcs) represent statistically meaningful relations and dependencies between these variables. When inferring a BN from real data, the network inference algorithm aims to recognize a model that is as close as possible to the observations made. As an effective solution, the BNs have been applied to several issues in power systems, such as fault diagnosis and reliability assessment.[14–16]

Here, the BN represents relationships among main variables in an AGC system, where variables can represent system frequency variation, disturbance, tie-line power deviation, load change, area control error, and the required control action command. In this work to model the AGC system, the measured/estimated variables are analyzed to produce an appropriate control signal. The probabilistic nature of BNs enables them to extract signals from noisy data and to naturally handle uncertainty that arises in the modeling of the AGC system. This probabilistic approach determines dependencies and conditional independencies among numerous variables, which is why it is able to include edges that represent meaningful relationships while excluding those edges that are not important.

8.1.1 BNs at a Glance

Sequential data arise in many areas of science and engineering. The data may be a time series, generated by a dynamical system, or a sequence generated by a one-dimensional spatial process. In such problems, it is desirable to find the probability of future outcomes as a function of our inputs, and using the BNs shows a way do that. The BNs as a form of flexible and interpretable (graphical) models have been proffered as a promising framework for modeling complex systems such as power systems, as they can represent probabilistic dependence relationships among multiple interacting components and variables. The BN models illustrate the effects of system components upon each other in the form of an influence diagram. These models can be automatically derived from experimental data through a statistically founded computational procedure known as network inference. Although the relationships are statistical in nature, they can sometimes be interpreted as causal influence connections.

As mentioned, the BNs are a form of graphical modeling, in which dependencies among variables are described in a graph including nodes and arcs (edges). The nodes represent variables, and the arcs represent dependencies. Dependencies are statistical in nature. Figure 8.1 shows a simple BN with variables A, B, and C. Variables B and C are statistically dependent upon variable A. Therefore, an edge from A to B in Figure 8.1 indicates that knowing A can help to predict B. This may or may not indicate a causal relationship, i.e., one in which A directly or indirectly affects B.

The BNs are able to handle incomplete data sets. For example, consider a problem with two strongly anticorrelated input variables. Since they cannot encode the correlation between the input variables, when one of the inputs is not observable, producing an accurate prediction using most existing models is impossible. The BNs offer a natural way to encode such dependencies.

The BNs allow one to learn about the causal relationships between different variables. They allow making predictions in the presence of interventions. The BNs, in conjunction with Bayesian statistical techniques, facilitate the combination of domain knowledge and data. They have a causal semantics that makes the encoding of causal prior knowledge particularly straightforward. The Bayesian methods in conjunction with the BNs and other types of models offer an efficient and principled approach for avoiding the overfitting of data, while there is no need to hold out some of the available data for testing purposes. In other words, using the Bayesian approach, the study models can be smoothed in such a way that all available data can be used for training.

The BN approaches are not model based and can be easily scalable for large-scale systems, such as real-world power systems. They can also work

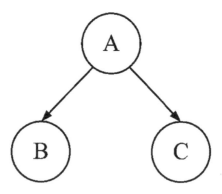

FIGURE 8.1
A BN depicting statistical dependencies among variables.

well in a nonlinear environment with variable structures. A major advantage of the BNs over many other types of predictive and learning models, such as neural networks, is their possibility of representing the interrelationships among the data set attributes.

In general, a BN consists of (1) an acyclic graph S, (2) a set of random variables $x = \{x_1, \ldots, x_n\}$ (the graph nodes) and a set of arcs that determine the nodes' (random variables') dependencies, and (3) a *conditional probability table* (CPT) associated with each variable ($p(x_i|pa_i)$). Together, these components define the joint probability distribution for x. The nodes in S are in one-to-one correspondence with the variables x. In this structure, x_i denotes a variable and its corresponding node, and pa_i represents the parents of node x_i in S as well as the variables corresponding to those parents. The lack of possible arcs in S encodes conditional independencies. In particular, given structure S, the joint probability distribution for x is defined by

$$p(x_1, \cdots, x_n) = \prod_{i=1}^{n} p(x_i|pa_i)$$

(8.1)

The probability encoded by a BN may be Bayesian or physical. In the case of building BNs from prior knowledge alone, the probabilities will be Bayesian, while learning these networks from data provides physical probabilities. The basic tasks related to the BNs are (1) the structure learning phase, i.e., finding the graphical model structure; (2) the parameter learning phase, i.e., finding nodes' probability distribution; and (3) the BN inference.

The structure and parameter learning are based on the prior knowledge and prior data (training data) of the model. The basic inference task of a BN consists of computing the posterior probability distribution on a set of query variables q, given the observation of another set of variables e called the evidence (i.e., $p(q|e)$). Different classes of algorithms have been developed to compute the marginal posterior probability $p(x|e)$ for each variable x, given the evidence e. No need to learn the inference data is an important characteristic in the BNs. Inference is a probabilistic action that obtains the probability of the query using prior probability distribution.

A *graphical model* to encode a set of conditional independence assumptions and a compact way of representing a joint probability distribution between random variables are the main features in a BN construction.

8.1.2 Graphical Models and Representation

In a BN, probabilistic relationships are usually represented by a simple directed acyclic graph (g). In the graph, the nodes represent variables and the edges represent dependencies. Therefore, the lake of edge between two variables indicates a conditional independency.[13] In a *directed acyclic graph*

(DAG), the edges are single-ended arrows, originating from one node (parent node, e.g., A in Figure 8.1) and ending in another (child node, e.g., B and C in Figure 8.1). Furthermore, a DAG does not include directed cycles, e.g., feedback from a node to itself. The type and value of dependency for each variable to its parent(s) is quantitatively described via a *conditioned probability distribution* (CPD). The CPD, which is consistent with the conditional independencies implied by g, can be described by a vector of its parameters. The variables in a BN may be continuous or discrete. However, in this chapter, for the AGC design issue, discrete variables are used. A BN represents the joint probability distribution for a finite set of random variables x, where $x_i \in x$ may take on a value from a specific domain.

Graphical models are generated by probability and graph theories to introduce a natural tool for dealing with two problems that occur throughout applied mathematics and engineering. In particular, they play a significant role in the synthesis and analysis of machine learning algorithms. Building a complex system by combining simpler parts is the fundamental idea of a graphical model. The probability theory provides the glue whereby the parts are combined, ensuring that the system as a whole is consistent, and providing ways to interface models to data. The graph theoretic side of graphical models provides both an intuitively appealing interface by which humans can model highly interacting sets of variables, and a data structure that lends itself naturally to the design of efficient general purpose algorithms.[21]

The graphical model formulation provides a natural framework for the design of new systems. Many of the classical multivariate probabilistic systems studied in the fields of engineering and science are special cases of the general graphical model formulation. The graphical model framework provides a way to view real-world systems by transferring specialized techniques that have been developed in one field to other areas.

As mentioned, in the probabilistic graphical models, the nodes represent random variables and the arcs represent conditional independence assumptions between variables. The arcs' pattern presents the graph structure. Hence, it provides a compact representation of joint probability distributions. For example, for N binary random variables, an atomic representation of the joint $p(x_1, \ldots, x_n)$ needs $O(2^n)$ parameters, whereas a graphical model may need exponentially fewer, depending on which conditional assumptions are considered.[21]

Usually, to construct a graphical model, the following two steps should be considered: (1) the model definition or learning phase for recognizing random variables (as graphical model nodes), the nodes' probability distribution estimation, and the nodes' dependencies identification (as model arcs), and (2) model inference for computing the posterior probability distribution on a set of query variables, given the observation of another set of variables (i.e., $p(q|e)$). Graphical models can be classified as undirected and directed models. The undirected graphical models (or Markov networks) are more suitable in the physical applications, while the directed graphical models (or BNs) are more popular with the AI and machine learning processes.

8.1.3 A Graphical Model Example

For the sake of illustration, consider an example on the blackout phenomenon following a significant disturbance/fault in a power system. In the event of a significant disturbance, generators and prime mover controls become important, as well as system control loops and special protections. In the case of improper coordination, it is possible for the system to become unstable, and generating units or loads may ultimately be tripped, possibly leading to a system blackout. As a power system fails, because of dynamic complexity and its high-order multivariable structure, more than one source of instability (frequency, angle, and voltage) may ultimately emerge. In a power system, the dynamic performance is influenced by a wide array of devices with different responses and characteristics. Hence, instability in a power system may occur in many different ways, depending on the system topology, operating mode, and form of the disturbance.

The impacts of a disturbance may also involve much of the system, depending on the power system topology, form of the disturbance, and operating mode. For example, a disturbance on a critical part, followed by its isolation by protective relays, will cause fluctuation in tie-line power flows, rotor angle speeds, and bus voltages. The machine speed variations will actuate prime mover governors and system frequency, the voltage deviation will affect both generator and voltage regulators, and the frequency and voltage deviation will affect the system loads to varying degrees, depending on their responses and different characteristics.[22] Furthermore, the various protective relays and devices may respond to these variations, and thereby affect the power system performance, possibly even leading to a system blackout.

Based on the above description, the graph shown in Figure 8.2 can be considered an example to explain the consequence of events leading to a blackout. One may represent the dependencies among more important variables in a BN graph as shown in Figure 8.2. This BN represents the joint probability distribution between a serious *disturbance* (D), *frequency* deviation (F), rotor *angle* variation (A), system *voltage* deviation (V), *protective* devices and *load* response (PL), and system *blackout* (B).

In this graph, A and V are the children of D, F is the child of A and D, PL is the child of F, A, and V, and B is the child of PL. D has no parents and is called a *root node*. Assume each variable can take on the value 1 (in the case of existing) or 0 (in the case of absenting).

While the graph appears to represent variable dependencies, its primary purpose is actually to encode conditional independencies, critical for their ability to confer a compact representation to a joint probability distribution.[23] In the above example, B is dependent upon PL, PL is dependent upon F, A, and V, A and V are dependent upon D, and F is dependent upon D and A.

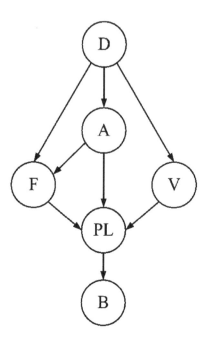

FIGURE 8.2
A BN structure example for main variables causing a blackout.

However, when the system is stabilized following a fault, B becomes independent of F, A, and V, as well as D. If we already know that a frequency deviation (voltage or angle variation) is occurring at $F = 1$ ($V = 1$, $A = 1$), then knowing something about the presence of D will not help us to determine the value of B as well as PL. Therefore, B and PL are conditionally independent of D given F, A, and V. Formally, x is conditionally independent of y given z if

$$p(x|y,z) = p(x|z) \tag{8.2}$$

The graph (g) encodes the Markov assumptions. As a consequence of the Markov assumption, the joint probability distribution over the variables represented by the BN can be factored into a product over variables, where each term is local conditional probability distribution of that variable, conditional on its parent variables (Equation 8.1).

The above chain rule states that the joint probability of independent entities is the product of their individual probabilities. A main advantage of the BN is its compact representation of the joint probability distribution. With no independence assumption, the joint probability distribution over the variables D, V, A, F, PL, and B is

$$P\,(D,V,A,F,PL,B) = P(D)\,P(V|D)\,P(A|D,V)\,P(F|D,V,A)\,P(PL|D,V,A,F)$$
$$P(B|D,V,A,F,PL) \tag{8.3}$$

For binary variables, this is $\sum_{i=1}^{6} 2^{i-1} = 1 + 2 + 4 + 8 + 16 + 32 = 63$ parameters. Employing the conditional independencies, the joint probability distribution for the graphical model shown in Figure 8.2 becomes

$$P(D, V, A, F, PL, B) = P(D) \, P(V|D) \, P(A|D) \, P(F|D, A) \, P(PL|V, A, F)$$
$$P(B|PL) \tag{8.4}$$

where, for binary variables of the example at hand, it is equal to $1 + 2 + 2 + 4 + 8 + 2 = 19$ parameters.

In the case of the multinomial distributions, each CPD can be presented in a CPT. For binary variables, in order to specify all the parameters of the CPD for variable x_i with parent(s) pa_i, each CPT must have 2^k entries, where k is the number of parents. Returning to the blackout example, it is assumed that the variables are binary. The parameters of the CPTs are shown in Figure 8.3.

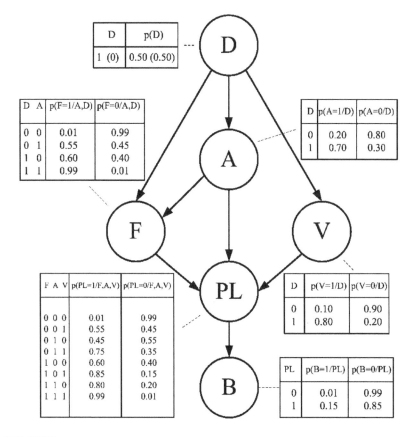

FIGURE 8.3
The BN for a blackout example with assumed CPTs.

The previous CPTs include nineteen parameters that are specified for the present BN. While the graph reveals the conditional independencies, the CPTs demonstrate the strength of dependencies. The previous tables indicate various probabilities; for instance, the probability of a disturbance (D) occurring is 0.5, the probability of having a frequency deviation (F) when a disturbance (D) and angle deviation (A) are present is 0.99, and the probability of a blackout after affecting the system load and the response of protective devices (PL) is 0.15. The tables show that although both D and A can cause F, D alone has a stronger effect than A alone:

$$P(F|D) = P(F = 1|D = 1, A = 0) = 0.6$$

$$P(F|A) = P(F = 1|D = 0, A = 1) = 0.55$$

In the above calculations, the third variable is fixed at zero. Ignoring this assumption, the results will be computed as follows, which shows a different result:

$$P(F \mid D) = P(F = 1 \mid D = 1)$$
$$= \frac{P(F = 1, D = 1)}{P(D = 1)} = \frac{P(F = 1, D = 1, A = 0) + P(F = 1, D = 1, A = 1)}{P(D = 1)}$$
$$= \frac{0.09 + 0.3465}{0.5} = 0.873$$

$$P(F \mid A) = P(F = 1 \mid A = 1)$$
$$= \frac{P(F = 1, A = 1)}{P(A = 1)} = \frac{P(F = 1, A = 1, D = 0) + P(F = 1, A = 1, D = 1)}{P(A = 1, D = 1) + P(A = 1, D = 0)}$$
$$= \frac{0.3465 + 0.055}{0.35 + 0.1} = 0.8922$$

8.1.4 Inference

Inference and reasoning from evidence and factual knowledge is the most common task in the BN applications, even for incomplete information. In an inference task, usually it is desirable to know the value of a particular node, without accessing that information. Usually, for this purpose, the evidence is in use. For example, in the previous example it is desirable to know if the disturbance (D) happened when the system frequency deviation (F) appeared (P(D|F)). Here, the F variable is called evidence. When evidence is available, one may reason about a cause of the instantiated variable. Bayes' rule is used to compute such a probability:

$$P(D|F) = \frac{P(D,F)}{P(F)} = \frac{P(F|D)\,P(D)}{p(F)} \tag{8.5}$$

Here, D is the unknown (hidden) node and F is the observed evidence. Without evidence, it is necessary to sum the joint probability distribution over all possible values of the other variables. For the given example, Equation 8.5 is calculated as follows:

$$P(D=1|F=1) = \frac{P(D=1, F=1)}{P(F=1)}$$

$$= \frac{\sum_A P(D=1, F=1, A)}{\sum_{A,D} P(F=1, A, D)}$$

$$= \frac{\begin{array}{c} P(D=1, F=1, A=0) \\ +P(D=1, F=1, A=1) \end{array}}{\begin{array}{c} P(F=1, D=1, A=1) + P(F=1, D=0, A=0) \\ +P(F=1, D=1, A=0) + P(F=1, D=0, A=1) \end{array}}$$

$$= \frac{0.4365}{0.4955} = 0.8809$$

and for $P(D = 0|F = 1)$,

$$P(D = 0|F = 1) = 1 - P(D = 1|F = 1) = 0.1191$$

As another example, it is desirable to know if the frequency deviation (F) happened when the system load and protective devices were affected (P(F|PL)). This probability can be computed as follows:

$$P(F=1|PL=1) = \frac{P(F=1, PL=1)}{P(PL=1)} = \frac{0.449413}{0.525292} = 0.855549$$

Therefore,

$$P(F = 0|PL = 1) = 1 - 0.855549 = 0.144451$$

Similarly, the probabilities for $P(F = 1|PL = 0)$ and $P(F = 0|PL = 0)$ can be calculated:

$$P(F = 1|PL = 0) = \frac{P(F = 1, PL = 0)}{P(PL = 0)} = \frac{0.046087}{0.474708} = 0.130619$$

$$P(F = 0|PL = 0) = 1 - 0.130619 = 0.869381$$

It is noteworthy that even for the binary case, the joint probability distribution has size $O(2^n)$, where n is the number of nodes. The full summation/integration over the joint probability distribution using discrete/continuous variables is called *exact inference*, and for large networks takes a long time. In response to this challenge, *approximate inference* methods have been proposed in the literature, such as the *Monte Carlo sampling* method (which provides gradually improving estimates as sampling proceeds[24]) and *Markov chain Monte Carlo* (MCMC) methods (including the *Gibbs sampling* and *Metropolis–Hastings algorithm*[25]).

8.1.5 Learning

In practical applications, the BN is usually unknown, and it is necessary to learn it from performed training data, expert knowledge, and other prior information. *Learning* in a BN can refer to the BN structure (graph topology) or the BN parameters of the related joint probability distribution. Over the years, several learning methods/algorithms have been developed, such as MCMC, *expectation maximization* (EM), *maximum likelihood estimation* (MLE), and *local search* (LS) through model space techniques.[17] Based on the BN structure (known/unknown) and rate of observability (full/partial), the existing BN learning methods can be classified into four classes, as shown in Table 8.1.

TABLE 8.1

Classes of BN Learning

Class	Structure	Observability	Learning Method
I	Known	Full	MLE
II	Known	Partial	EM, MCMC
III	Unknown	Full	LS
IV	Unknown	Partial	EM, LS

8.2 AGC with Wind Farms

8.2.1 Frequency Control and Wind Turbines

The dynamic behavior of a power system in the presence of wind power units might be different from that in conventional power plants.[7] The power outputs of such sources are dependent on weather conditions, seasons, and geographic location. When wind power is a part of a power system, additional imbalance is created when the actual wind power deviates from its forecast due to wind velocity variations (see Chapter 6). So, scheduling conventional generator units to follow the load (based on the forecasts) may also be affected by wind power output.

Furthermore, the effect of wind farms on the dynamic behavior of a power system may cause a different system frequency response to a disturbance event (such as load disturbance). Since the system inertia determines the sensitivity of the overall system frequency, it plays an important role in this consideration. Lower system inertia leads to faster changes in the system frequency following a load-generation variation. The addition of synchronous generation to a power system intrinsically increases the system inertial response.

In practice, an AGC system must be designed to maintain the system frequency and tie-line power deviations within the limits of specified frequency operating standards. In the existing standards, the acceptable range of frequency deviation for near-normal operation (AGC operating area) is small. This range is mainly determined by the available amount of operating reserved power in the system. It is noteworthy that the total AGC power reserve in a real power system is usually about 15%, and the AGC loop can usually track the load variation in the range of 1 to 10% of the overall load demand.

The AGC system is designed to operate during a relatively small and slow change in real power load and frequency. For large imbalances in real power associated with rapid frequency changes that occur during a fault condition, the AGC system is unable to restore the frequency. There is a risk that these large frequency deviation events might be followed by additional generation events, load/network events, separation events, or multiple contingency events. For such large frequency deviations and in a more complex condition, the emergency control and protection schemes must be used to restore the system frequency.[1]

This intrinsic increase does not necessarily occur with the addition of wind turbine generators (WTGs) due to their differing electromechanical characteristics.[4] So, the impact of wind farms on power system inertia is a key factor in investigating the power system frequency behavior in the presence of a high penetration of wind power generation.

As mentioned in Chapter 6, to analyze the additional variation caused by WTGs, the total effect is important, and every change in wind power output does not need to be matched one for one by a change in another generating unit moving in the opposite direction. The slow WTG power fluctuation dynamics and total average power variation negatively contribute to the power imbalance and frequency deviation, which should be taken into account in the well-known AGC control scheme. This power fluctuation must be included in the conventional area control error (ACE) formulation.

8.2.2 Generalized ACE Signal

The conventional AGC model is discussed in Chapter 2, Bevrani[1] and Kundur.[26] Following a load disturbance within the control area, the frequency of the area experiences a transient change and the feedback mechanism generates an appropriate rise or lower signal in the participating generator units according to their participation factors to make the generation follow the load. In the steady state, the generation is matched with the load, driving the tie-line power and frequency deviations to zero. As there are many conventional generators in each area, the control signal has to be distributed among them in proportion to their participation.[27]

The frequency performance of a control area is represented approximately by a lumped load-generation model using equivalent frequency, inertia, and damping factors.[28] Because of the range of use and specific dynamic characteristics, such as a considerable amount of kinetic energy, the wind units are more important than the other renewable energy resources. The equivalent system inertia in a power system with wind units can be defined as

$$H = H_{sys} = H_C + H_W = \sum_{i=1}^{N1} H_{Ci} + \sum_{i=1}^{N2} H_{Wi} \tag{8.6}$$

where H_{sys} is equivalent to the inertia constant. H_C and H_W are the total inertia constants due to conventional and wind turbine generators, respectively. The inertia constant for wind power is time dependent. The typical inertia constant for the wind turbines is about 2 to 6 s.[1]

Similar to the given analysis and problem formulation in Section 6.1, the updated ACE signal should represent the impacts of wind power on the scheduled flow over the tie-line. The ACE signal is traditionally defined as a linear combination of frequency and tie-line power changes as follows:[26]

$$ACE = \beta \Delta f + \Delta P_{tie} \tag{8.7}$$

where Δf is frequency deviation, β is frequency bias, and ΔP_{tie} is the difference between the actual (*act*) and scheduled (*sched*) power flows over the tie-lines.

$$\Delta P_{tie} = \sum (P_{tie,act} - P_{tie,sched})$$ (8.8)

For a considerable amount of wind (W) power, its impact must also be considered with conventional (C) power flow in the overall area tie-line power. Therefore, the updated ΔP_{tie} can be expressed as follows:

$$\Delta P_{tie} = \Delta P_{tie-C} + \Delta P_{tie-W}$$
$$= \sum (P_{tie-C,act} - P_{tie-C,sched}) + \sum (P_{tie-W,act} - P_{tie-W,estim})$$ (8.9)

The total wind power flow change is usually smooth compared to variation impacts from the individual wind turbine units. Using Equations 8.7 and 8.9, an updated ACE signal can be completed as

$$ACE = \beta \Delta f + \sum (P_{tie-C,act} - P_{tie-C,sched})$$
$$+ \sum (P_{tie-W,act} - P_{tie-W,estim})$$ (8.10)

where $P_{tie-C,act}$, $P_{tie-C,sched}$, $P_{tie-W,act}$, and $P_{tie-W,estim}$ are actual conventional tie-line power, scheduled conventional tie-line power, actual wind tie-line power, and estimated wind tie-line power, respectively.

In typical AGC implementations, the system frequency gradient and ACE signal must be filtered to remove noise effects before use. The ACE signal then is often applied to the controller block. Control dead-band and ramping rate are different for various systems.[1] The controller can send higher or lower pulses to generating plants if its ACE signal exceeds a standard limit.

8.3 Proposed Intelligent Control Scheme

8.3.1 Control Framework

The overall view of the proposed control framework for a typical area i is conceptually shown in Figure 8.4. The control scheme in each area presents two agents: an estimator agent for the estimation amount of load change, and an intelligent BN-based controller agent to provide an appropriate supplementary control action signal. The objective of the proposed design is to regulate the frequency and tie-line power in the power system concerning the integration of wind power units with various load disturbances and achieve a desirable control performance.

The two-agent schema entailed the minimum number of measurement/ monitoring and control activities in a control area to track the AGC tasks.

Measured Data

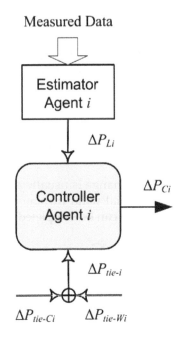

FIGURE 8.4
The proposed BN-based multiagent control for area i.

The controller agent uses ΔP_{tie} (Equation 8.9) and load demand change (ΔP_L) signals to provide the control action signal (ΔP_C). The estimator agent is responsible for estimating the amount of load change.

8.3.2 BN Structure

To find a clear view of the BN structure, it is better to start by determining the necessary variables for modeling. This initial task is not always straight-forward. As part of this task, one should (1) correctly identify the modeling objective, (2) investigate important relevant observations, (3) determine what subset of those observations is worthwhile to model, and finally, (4) organize the observations into variables having mutually exclusive and collectively exhaustive states.

In the process of a BN construction for the AGC issue, the aim is to achieve the AGC objective and keep the ACE signal within a small band around zero using the supplementary control action signal. Then, the query variable in the posterior probability distribution is a ΔP_C signal and the posterior probabilities according to possible observations relevant to the problem (as shown in Figure 8.5) are as follows:

$$p(\Delta P_C | ACE, \Delta P_{tie}, \Delta P_L, \Delta f)$$

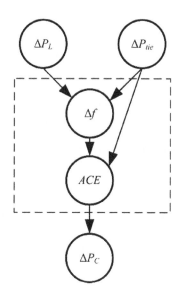

FIGURE 8.5
The graphical model for an area.

$$p(\Delta P_C | ACE, \Delta P_L, \Delta f)$$

$$p(\Delta P_C | \Delta P_{tie}, \Delta P_L, \Delta f)$$

$$p(\Delta P_C | ACE, \Delta P_{tie}, \Delta P_L)$$

$$p(\Delta P_C | ACE, \Delta f)$$ (8.11)

$$p(\Delta P_C | \Delta P_{tie}, \Delta P_L)$$

$$\vdots$$

$$p(\Delta P_C | \Delta P_{tie})$$

$$p(\Delta P_C | \Delta P_L)$$

According to Equation 8.11, there are many observations that are related to the AGC problem; however, the best ones, which have the least dependency on the model parameters (e.g., frequency bias factor, etc.) and cause the maximum impact on the frequency deviation, and consequently ACE signal changes, are load disturbance and tie-line power deviation signals. Then the appropriate posterior probability that should be found is $p(\Delta P_C | \Delta P_{tie}, \Delta P_L)$.

The ΔP_{tie} can be practically obtained using measurement instruments. However, ΔP_L is one of the input parameters that is not measurable directly, but it can be easily estimated using a numerical/analytical method. A simple method to estimate the amount of load change following a load disturbance is discussed in the next section. This estimation method is initially based on the measured frequency gradient and the specified system characteristics. Considering the AGC duty cycle (timescale), the total consumed time needed for the estimation process is not significant.

Another approach for constructing the graphical model of a BN can be described based on the following observations:[29] From the chain rule of probability, it is known that

$$p(x) = \prod_{i=1}^{n} p(x_i | x_1, \cdots, x_{i-1}) \tag{8.12}$$

For every x_i, there will be some subset $\pi_i \subseteq \{x_1, \ldots, x_{i-1}\}$ such that x_i and $\{x_1, \ldots, x_{i-1}\}/\pi_i$ are conditionally independent given π_i. That is, for any x,

$$p(x_i | x_1, \cdots, x_{i-1}) = p(x_i | \pi_i) \tag{8.13}$$

Combining Equations 8.12 and 8.13,

$$p(x) = \prod_{i=1}^{n} p(x_i | \pi_i) \tag{8.14}$$

Comparing Equations 8.8 and 8.10 shows that the variables sets $(\pi_1, \ldots, \pi_n)$ correspond to the BN parents $(pa_1, \ldots, pa_n)$, which in turn fully specify the arcs in the network structure S. Consequently, to determine the structure of a BN, one must (1) order the variables somehow and (2) determine the variables sets that satisfy Equation 8.13 for $i = 1, \ldots, n$.

8.3.3 Estimation of Amount of Load Change

As mentioned, the estimator agent estimates the total power imbalance (amount of load change sensed in control area ΔP_L) by an assigned algorithm based on the following explained analytical method.

Consider the ith generator swing equation for a control area with N generators ($i = 1, \ldots, N$):

$$2H_i \frac{d\Delta f_i(t)}{dt} + D_i \Delta f_i(t) = \Delta P_{mi}(t) - \Delta P_{Li}(t) = \Delta P_{di} \tag{8.15}$$

where ΔP_{mi} is the mechanical turbine power, ΔP_{Li} is the load demand (electrical power), H_i is the inertia constant, D_i is load damping, and ΔP_{di} is the load-generation imbalance. By adding N generators within the control area, one obtains the following expression for the total load-generation imbalance:

$$\Delta P_D(t) = \sum_{i=1}^{N} \Delta P_{di}(t) = 2H \frac{d\Delta f(t)}{dt} + D\Delta f(t) \tag{8.16}$$

Equation 8.16 shows the multimachine dynamic behavior by an equivalent single machine. Using the concept of an equivalent single machine, a control area can be represented by a lumped load-generation model using an equivalent frequency Δf, system inertia H, and system load damping D.[1]

$$\Delta f = \Delta f_{sys} = \sum_{i=1}^{N} (H_i \Delta f_i) / \sum_{i=1}^{N} H_i \tag{8.17}$$

$$H = H_{sys} = \sum_{i=1}^{N} H_i, \quad D = D_{sys} = \sum_{i=1}^{N} D_i \tag{8.18}$$

The magnitude of the total load-generation imbalance ΔP_D, after a while, can be obtained from Equation 8.16.

$$\Delta P_D = D\Delta f \tag{8.19}$$

where

$$\Delta P_D = \Delta P_m - \Delta P_L - \Delta P_{tie} \tag{8.20}$$

$$\Delta P_m = \sum_{i=1}^{N} P_{mi} \tag{8.21}$$

$$\Delta P_L = \sum_{i=1}^{N} P_{Li} \tag{8.22}$$

and ΔP_{tie} is defined in Equation 8.9. The total mechanical power change indicates the total power generation change due to governor action, which is in proportion to the system frequency deviation:[26]

$$\Delta P_m \cong \Delta P_g = -\frac{1}{R_{sys}} \Delta f \tag{8.23}$$

Equations 8.19, 8.20, and 8.23 give

$$\Delta P_L = -\left(\frac{1}{R_{sys}} + D\right)\Delta f - \Delta P_{tie} \tag{8.24}$$

Thus, the total load change in a control area is proportional to the system frequency deviation. Neglecting the network power losses, $\Delta P_D(t)$ shows the load-generation imbalance proportional to the total load change. Using Equation 8.16, the magnitude of total load-generation imbalance immediately after the occurrence of disturbance at $t = 0^+s$ can be expressed as follows:

$$\Delta P_D = 2H_{Sys} \frac{d\Delta f}{dt} \tag{8.25}$$

Equations 8.20 and 8.25 give

$$\Delta P_L = \Delta P_m - 2H_{Sys}\frac{d\Delta f}{dt} - \Delta P_{tie} \tag{8.26}$$

Here, Δf is the frequency of the equivalent system. To express the result in a suitable form for sampled data, Equation 8.26 can be represented in the following difference equation:

$$\Delta P_L(T_S) = \Delta P_m(T_S) - \frac{2H_{Sys}}{T_S}[\Delta f_1 - \Delta f_0] - \Delta P_{tie}(T_S) \tag{8.27}$$

where T_S is the sampling period and Δf_1 and Δf_0 are the system equivalent frequencies at t_0 and t_1 (the boundary samples within the assumed interval).

8.4 Implementation Methodology

8.4.1 BN Construction

After determining the most worthwhile subset of the observations $(\Delta P_{tie}, \Delta P_L)$, in the next phase of the BN construction, a directed acyclic graph that encodes assertion of conditional independence is built. It includes the problem random variables, conditional probability distribution nodes, and dependency nodes.

The basic structure of the needed graphical model for the AGC issue, which is shown in Figure 8.5, is built based on prior knowledge of the problem. According to Equation 8.7, the ACE signal is dependent on the frequency and tie-line power deviations; then they will be the parent nodes of the ACE signal (control input) in the BN graphical model, and since frequency deviation is dependent on the load disturbance and tie-line power deviation, they will be parent nodes of Δf. Since ΔP_C as the controller output is considered to be dependent on the ACE signal only, the ACE node will be the parent node for the control action signal.

Using the ordering $(\Delta P_{tie}, \Delta P_L, \Delta f, ACE, \Delta P_C)$ and according to Equation 8.13, the conditional dependencies are described as follows:

$$p(\Delta P_L | \Delta P_{tie}) = p(\Delta P_L)$$

$$p(\Delta P_{tie} | \Delta P_L) = p(\Delta P_{tie})$$

$$p(\Delta f | \Delta P_L, \Delta P_{tie}) = p(\Delta f | \Delta P_L, \Delta P_{tie}) \qquad (8.28)$$

$$p(ACE | \Delta P_{tie}, \Delta P_L, \Delta f) = p(ACE | \Delta P_{tie}, \Delta f)$$

$$p(\Delta P_C | ACE, \Delta P_{tie}, \Delta P_L, \Delta f) = p(\Delta P_C | ACE)$$

In the graphical model of a BN, each node presents a system variable. The edges between nodes present dependency between the system variables. In a BN, the aim is to find the probability distribution of the graphical model nodes from training data (parameter learning), and then do an inference task according to the observation. In the graphical model each node has a probability table and nodes with parents have conditional probability tables (because they are dependent on their parents).

The graphical model of the AGC problem (Figure 8.5) is based on the right side of the described relationships in Equation 8.28. In the next step of the BN construction (parameter learning), the local conditional probability distributions $p(x_i|pa_i)$ must be computed from the training data. Probability and conditional probability distributions related to the AGC issue, according to Figure 8.5, are $p(\Delta P_L)$, $p(\Delta P_{tie})$, $p(\Delta f|\Delta P_L,\Delta P_{tie})$, $p(ACE|\Delta P_{tie},\Delta f)$, and $p(\Delta P_C|ACE)$. To find these probabilities, the training data matrix should be provided.

Here, the Bayesian networks toolbox (BNT)[17] is used for probabilistic inference of the model. The BNT uses the training data matrix and finds the conditional probabilities related to the graphical model variables (as the parameter learning phase).

Once a BN is constructed (from prior knowledge, data, or a combination), various probabilities of interest from the model can be determined. For the problem at hand, it is desired to compute the posterior probability distribution on a set of query variables, given the observation of another set of variables called the evidence. The posterior probability that should be found is $p(\Delta P_C|\Delta P_{tie},\Delta P_L)$. This probability is not stored directly in the model, and hence needs to be computed. In general, the computation of a probability of interest given a model is known as *probabilistic inference*.

8.4.2 Parameter Learning

As mentioned and shown in the graphical model of a control area (Figure 8.5), the essential parameters used for the learning phase among each control area of a power system can be considered as ΔP_{tie}, ΔP_L, Δf, ACE, and ΔP_C.

In order to find a related set of training data (ΔP_{tie}, ΔP_L, Δf, ACE, ΔP_C) for the sake of the parameter learning phase, one can provide a long-term simulation for the considered power system case study in the presence of various disturbance scenarios. This large learning set is partly complete and can be used for the parameter learning issue in the power system with a wide range of disturbances. Since the BNs are based on inference and new cases (which may not be included in the training set) can be inferred from the training table data, it is not necessary to repeat the learning phase of the system for different amounts of disturbances occurring in the system.

After providing the training set, the training data related to control areas are given to the BNT separately. The BNT uses the input data and does the parameter learning phase for each control area's parameters. It finds prior and conditional probability distributions related to that area's parameters, which, according to Figure 8.5, are $p(\Delta P_L)$, $p(\Delta P_{tie})$, $p(\Delta f|\Delta P_L,\Delta P_{tie})$, $p(ACE|\Delta P_{tie},\Delta f)$, and $p(\Delta P_C|ACE)$. Following completion of the learning phase, the power system simulation will be ready to run and the proposed model uses the inference phase to find an appropriate control action signal (ΔP_C) for each control area.

During the simulation stage, the inference phase is done as follows: At each simulation time step, the corresponding controller agent of each area gets the input parameters $(\Delta P_{tie}, \Delta P_L)$ of the model and digitizes them for the BNT (the BNT does not work with continuous values). The BNT finds the posterior probability distribution values $p(\Delta P_C | \Delta P_{tie}, \Delta P_L)$ related to each area. Then, the controller agent finds the maximum posterior probability distribution from the return set, and gives the most probable evidence ΔP_C in the control area.

8.5 Application Results

In order to illustrate the effectiveness of the proposed intelligent control strategy, first we examine it on the well-known New England ten-generator, thirty-nine-bus system as a test case study. Then, we explain an experimental real-time application on a longitudinal four-machine infinite bus system. The results in both test systems are compared with the application of a multiagent reinforcement learning (RL) AGC approach, which was presented briefly in Chapter 7. Interested readers can find more detail on the AGC synthesis using the reinforcement learning technique in Daneshfar and Bevrani[30] and Bevrani et al.[31]

8.5.1 Thirty-Nine-Bus Test System

The thirty-nine-bus test system is widely used as a standard system for testing of new power system analysis and control synthesis methodologies, as well as in Chapters 6, 7, and 11 of this book. A single line diagram of the system is given in Figure 8.6. This system has ten generators, nineteen loads, thirty-four transmission lines, and twelve transformers. The well-known test system is updated by adding two wind farms in buses 5 and 21, as shown in Figure 8.6. The system is divided into three areas.

The total system installed capacity is 841.2 MW of conventional generation and 45.34 MW of wind power generation. There are 198.96 MW of conventional generation, 22.67 MW of wind power generation, and 265.25 MW load in area 1. In area 2, there are 232.83 MW of conventional generation and 232.83 MW of load. In area 3, there are 160.05 MW of conventional generation, 22.67 MW of wind power generation, and 124.78 MW of load.

The simulation parameters for the generators, loads, lines, and transformers of the test system are given in Bevrani et al.[31] All power plants in the power system are equipped with a speed governor and power system stabilizer (PSS). However, only one generator in each area is responsible for the AGC task: G1 in area 1, G9 in area 2, and G4 in area 3.

In the present work, similar to the real-world power systems, it is assumed that conventional generation units are responsible for providing spinning reserve for the purpose of load tracking and the AGC task. For the sake of

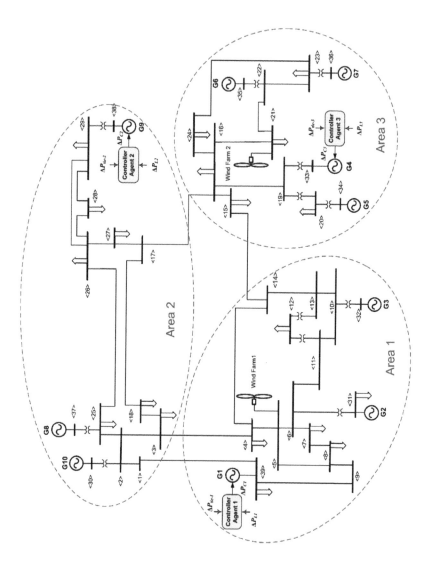

FIGURE 8.6
Single line diagram of thirty-nine-bus test system with wind farms.

simulation, random variations of wind velocity have been considered. The dynamics of WTGs, including the pitch angle control of the blades, are also considered. The start-up and rated wind velocities for the wind farms are specified as about 8.16 and 14 m/s, respectively. Furthermore, the pitch angle controls for the wind blades are activated only beyond the rated wind velocity. The pitch angles are fixed to 0° at the lower wind velocity below the rated one.

To cope with real-world power systems, in the performed application the important inherent requirement and basic constraints, such as governor dead-band and generation rate constraint imposed by physical system dynamics, are considered. For the sake of simulation, three step load disturbances are simultaneously applied to the three areas: 3.8% of the total area load at bus 8 in area 1, 4.3% of the total area load at bus 3 in area 2, and 6.4% of the total area load at bus 16 in area 3. Using the simulation, the training table rows can be built in the format shown in Table 8.2. The applied step load disturbances ΔP_{Li} (pu), the output power of wind farms P_{WT} (MW), and the wind velocity V_W (m/s) are considered as shown in Figure 7.5.

A simple presentation of probability tables using the proposed graphical model (Figure 8.5), according to the training data, after the parameter learning phase for the test system, is shown in Table 8.3. Some samples of returned posterior probability distribution values $p(\Delta P_C|\Delta P_{tie}, \Delta P_L)$ from BNT environment are also shown in Table 8.4.

The frequency deviation (Δf) and area control error (ACE) signals of the closed-loop system are shown in Figures 8.7 and 8.8. As a sample, the produced mechanical power by the AGC participant unit in area 2, corresponding electrical power, and overall tie-line power for the same area are shown in Figure 8.9.

In the proposed simulations, the wind power impacts on the overall system frequency behavior can be clearly seen. The fast movements in wind power output are combined with movements in load and other resources. The power system response is affected by the wind power fluctuation. When

TABLE 8.2

Training Data Matrix for Area i

Time (s)	$\Delta P_{tie\text{-}i}$ (pu)	$\Delta P_{L\text{-}i}$ (pu)	Δf_i (Hz)	ACE_i (pu)	$\Delta P_{C\text{-}i}$ (pu)
0.005	0.03	0.1	−0.01	0.02	−0.08
...	...	...	...	...	...

TABLE 8.3

Returned Posterior Probability Distribution Values from BNT of Area i

$p(\Delta P_{C\text{-}i}|\Delta P_{tie\text{-}i}, \Delta P_{L\text{-}i})$	0.005	0.1	0.032	0	...
$\Delta P_{C\text{-}i}$ (pu)	−0.08	0.03	0.1	−0.005	...

TABLE 8.4

Some Samples of Probabilities According to the Graphical Model

ΔP_{tie} (pu)		ΔP_L (pu)		Δf (Hz)			
ΔP_{tie}	$p(\Delta P_{tie})$	ΔP_L	$p(\Delta P_L)$	ΔP_{tie}	ΔP_L	Δf	$p(\Delta f \mid \Delta P_L, \Delta P_{tie})$
−0.03	0.14	0.01	0.75	−0.03	0.01	0.002	0.75
−0.1	0.28	−0.2	0.02	−0.1	−0.2	−0.04	0.02
...	...	...	...	...	...	...	...

		ACE (pu)			ΔP_C (pu)		
ΔP_{tie}	Δf	ACE	$p(ACE \mid \Delta P_{tie}, \Delta f)$	ACE	ΔP_C	$p(\Delta P_C \mid ACE)$	
−0.03	0.002	0.01	0.06	0.01	−0.04	0.7	
−0.1	−0.04	0.031	0.81	0.031	0.1	0.01	
...	...	...	...	...	...	...	

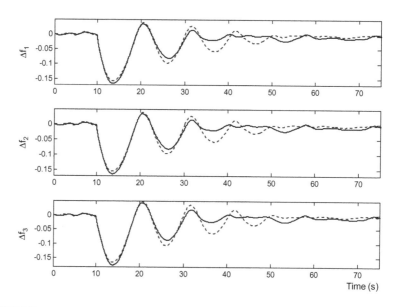

FIGURE 8.7
Frequency deviation: proposed multiagent BN method (solid line) and multiagent RL method (dashed line).

wind power is a part of the power system, an additional imbalance is created when the actual wind output deviates from its forecast. Before the load disturbance occurred in 10 s, there was also a little oscillation during the simulation that was initially caused from random variations of wind velocity, and it was magnified due to the resulting power fluctuation.

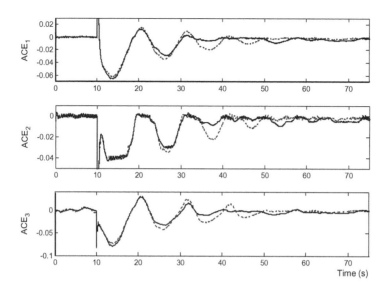

FIGURE 8.8
ACE signal: proposed multiagent BN method (solid line) and multiagent RL method (dashed line).

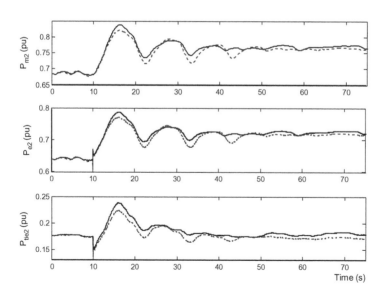

FIGURE 8.9
Area 2 responses: proposed multiagent BN method (solid line) and multiagent RL method (dashed line).

It is shown that using the proposed method, the area control error and frequency deviations in all areas are properly driven close to zero in the presence of wind turbines and load disturbance. Furthermore, the areas' frequency deviation is less than the frequency deviation in the system with multiagent RL-based controllers.

8.5.2 A Real-Time Laboratory Experiment

Since the AGC as a supplementary control is known as a long-term control problem (few seconds to several minutes), it is expected that the proposed AGC methodology will be successfully applicable in real-world power systems. To illustrate the capability of the proposed control strategy in real-time AGC applications, an experimental study has been performed on the large-scale Analog Power System Simulator (APSS)[1] at the Research Laboratory of the Kyushu Electric Power Company (Japan). For the purpose of this study, a longitudinal four-machine infinite bus system is considered as the test system. A single line diagram of the study system is shown in Figure 8.10a. All generator units are of the thermal type, with separately conventional excitation control systems. The set of four generators represents a control area (area I), and the infinite bus is considered as the other connected systems (area II).

The detailed information of the system, the parameters of generator units, and turbine systems (including the high-pressure, intermediate-pressure, and low-pressure parts) are given in Bevrani and Hiyama.[28] The wind farm (WF) consists of 200 units of 2 MW rated variable speed wind turbines (VSWTs). The VSWT parameters are indicated in Table 8.5. Although in the given model the number of generators is reduced to four, it closely represents the dynamic behavior of the West Japan Power System. As described in Bevrani and Hiyama,[28] the most important global and local oscillation modes of the actual system are included.

The whole power system (shown in Figure 8.10a) has been implemented using the APSS. Figure 8.10b shows an overview of the applied laboratory experimental devices, including the generator panels, monitoring displays, and control desk. The proposed control scheme, including estimator and controller agents, was built in a personal computer and connected to the power system using a digital signal processing (DSP) board equipped with analog-to-digital (A/D) and digital-to-analog (D/A) converters. The converters act as the physical interfaces between the personal computer and the analog power system hardware.

The performance of the closed-loop system is tested in the presence of load disturbances. The nominal area load demands are also fixed at the same values given in Bevrani and Hiyama.[28] More than 10% of the total demand power is supplied by the installed wind farm in bus 9.

For the first scenario, the power system is tested following a step loss of 0.06 pu conventional generation. The participation factors for Gen 1, Gen 2,

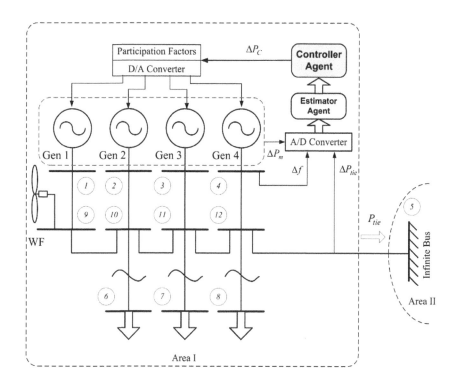

(a)

(b)

FIGURE 8.10
Performed laboratory experiment: (a) block diagram representation and (b) physical configuration.

TABLE 8.5

The VSWT Parameters

Wind speed (m/s)	11
Rotation speed (pu)	1.17
Inertia constant (s)	4.5
Magnetizing reactance (pu)	3.96545
Rotor reactance (pu)	0.1
Stator reactance (pu)	0.09273
Rotor resistance (pu)	0.00552
Stator resistance (pu)	0.00491

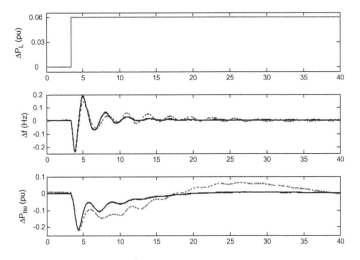

FIGURE 8.11

System response following a 0.06 pu step load change: proposed multiagent BN method (solid line) and multiagent RL method (dashed line).

Gen 3, and Gen 4 are fixed at 0.4, 0.25, 0.20, and 0.15, respectively. The applied step disturbance and the closed-loop system response, including frequency deviation (Δf) and tie-line power change (ΔP_{tie}), are shown in Figure 8.11. This figure shows that the frequency deviation and tie-line power change are properly maintained within a narrow band.

As a severe test scenario, the power system is examined in the presence of a sequence of step load changes. The load change pattern and the system response are shown in Figure 8.12. The obtained results show that the designed controllers can ensure good performance despite load disturbances. It is shown that the proposed intelligent AGC system acts to maintain area frequency and total exchange power closed to the scheduled values by sending a corrective smooth signal to the generators in proportion to their participation in the AGC task.

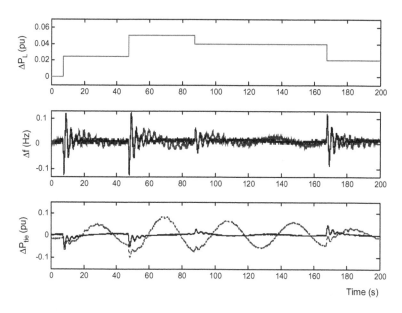

FIGURE 8.12

System response following a sequence of step load changes: proposed multiagent BN method (solid line) and multiagent RL method (dashed line).

The experiment results illustrate that the system performance using the proposed BN-based multiagent controller is quite better than the multiagent RL technique. It has been shown[30,31] that the multiagent RL design presents greater performance than the conventional (well-tuned) proportional-integral (PI)-based AGC systems. Therefore, in comparison with an existing conventional AGC system, the closed-loop performance of the present BN-based multiagent scheme is significantly improved. In summary, flexibility, a higher degree of intelligence, model independency, and handling of incomplete measured data (uncertainty consideration) can be considered some important advantages of the proposed methodology.

In the real-time simulations, the detailed dynamic nonlinear models of wind turbines are used without applying an aggregation model for the turbine units. That is why in the simulation results, in addition to the long-term fluctuations, the faster dynamics on a timescale of seconds are also observable.

Although the nonlinear simulation results for the test systems illustrate the capability of the proposed intelligent-based AGC scheme, certainly the observed dynamic response is still far from a real-world large-scale power system, with tens of conventional power plants and numerous distributed wind generators and other power sources. In a real large-scale power system with a high penetration of real power, the power system does not need to respond to the variability of each individual wind turbine.[8] Rapidly varying components of system signals are almost unobservable due to various filters involved in the process.

8.6 Summary

A new intelligent methodology for AGC synthesis concerning the integration of wind power units, using an agent-based Bayesian network, has been proposed for a large-scale power system. The proposed method was applied to the ten-generator, thirty-nine-bus power system as a test case study. An experimental examination was also performed on the APSS system.

The results show that in comparison with a designed agent-based reinforcement learning control, the new algorithm presents a desirable performance. Two important features of the new approach, i.e., model independence from power system model parameters and flexibility in specifying the control objectives, make it very attractive for frequency control practices. However, the scalability of BN-based systems for realistic problem sizes is one of the great reasons to use it in AGC design. In addition to scalability and benefits owing to the distributed nature of the multiagent solution, such as parallel computation, the BNs provide a robust probabilistic method of reasoning with uncertainty. They are more suitable for representing complex dependencies among components and can take into consideration load uncertainty as well as dependency of load in different areas.

References

1. H. Bevrani. 2009. *Robust power system frequency control*. 1st ed. New York: Springer Press.
2. P. S. Dokopoulos, A. C. Saramourtsis, A. G. Bakirtzis. 1996. Prediction and evaluation of the performance of wind-diesel energy systems. *IEEE Trans. Energy Conversion* 11:385–93.
3. R. B. Chedid, S. H. Karaki, C. El-Chamali. 2000. Adaptive fuzzy control for wind diesel weak power systems. *IEEE Trans. Energy Conversion* 15:71–78.
4. G. Lalor, A. Mullane, M. O'Malley. 2005. Frequency control and wind turbine technologies. *IEEE Trans. Power Syst.* 20:1905–13.
5. J. Morren, S. W. H. de Haan, W. L. Kling, J. A. Ferreira. 2006. Wind turbines emulating inertia and supporting primary frequency control. *IEEE Trans. Power Syst.* 21:433–34.
6. H. Banakar, Ch. Luo, B. Teck Ooi. 2008. Impacts of wind power minute-to-minute variations on power system operation. *IEEE Trans. Power Syst.* 23:150–60.
7. PSERC. 2009. *Impact of increased DFIG wind penetration on power systems and markets*. Final project report. http://www.pserc.org/docsa/.
8. M. Milligan, B. Kirby, R. Gramlich, M. Goggin. 2009. *Impacts of electric industry structure on high wind penetration potential*. NREL Technical Report. http://www.nrel.gov/docs/fy09osti/46273.pdf.
9. T. Hiyama, et al. 2005. Experimental studies on multi-agent based AGC for isolated power system with dispersed power sources. *Eng. Intelligent Syst.* 13(2):135–40.

10. T. Hiyama, et al. 2004. Multi-agent based operation and control of isolated power system with dispersed power sources including new energy storage device. In *Proceedings of International Conference on Renewable Energies and Power Quality (ICREPO'04)*, CD-ROM.

11. J. Morel, H. Bevrani, T. Ishii, T. Hiyama. 2010. A robust control approach for primary frequency regulation through variable speed wind turbines. *IEEJ Trans. Power Syst. Energy*, 130(11): 1002–009.

12. H. Bevrani, A. Ghosh, G. Ledwich. 2010. Renewable energy sources and frequency regulation: Survey and new perspectives. *IET Renewable Power Gener.*, 4(5): 438–57.

13. J. Pearl. 1988. *Probabilistic reasoning in intelligent systems: Networks of plausible inference*. San Mateo, CA: Morgan Kaufmann Publishers.

14. Y. Zhu, H. Limin, J. Lu. 2006. Bayesian networks-based approach for power systems fault diagnosis. *IEEE Trans. Power Delivery* 21:634–39.

15. D. C. Yu, T. C. Nguyen, P. Haddawy. 1999. Bayesian network model for reliability assessment of power systems. *IEEE Trans. Power Syst.* 14:426–32.

16. C. F. Chien, S. L. Chen, Y. S. Lin. 2002. Using Bayesian network for fault location on distribution feeder. *IEEE Trans. Power Delivery* 17:785–93.

17. K. Murphy. 2001. The Bayes net toolbox for Matlab. *Comput. Sci. Stat.* 33: 1–20.

18. D. Heckerman. 1996. *A tutorial on learning with Bayesian networks*. Technical report. Redmond, WA: Microsoft Research, 301–54.

19. J. Pearl. 2000. *Causality: Models, reasoning, and inference*. Cambridge: Cambridge University Press.

20. A. Bobbio, L. Portinale, M. Minichino, E. Ciancamerla. 2001. Improving the analysis of dependable systems by mapping fault trees into Bayesian networks. *Reliability Eng. Syst. Safety J.* 71:249–60.

21. K. Murphy. 2001. *An introduction to graphical models*. Technical report. Intel Research. www.cs.ubc.ca/~murphyk

22. P. Kundur. 2007. Power system stability. In *Power system stability and control*, chap. 7. Boca Raton, FL: CRC.

23. K. Sachs. 2006. Bayesian network models of biological signaling pathways. PhD Dissertation, MIT.

24. J. Pearl. 1987. Evidential reasoning using stochastic simulation of causal models. *Artificial Intelligence* 32(2):245–58.

25. T. L. Griffiths, A. Yuille. 2006. A primer on probabilistic inference. *Trends Cognitive Sci. Suppl.* 10(7):1–11.

26. P. Kundur. 1994. *Power system stability and control*. Englewood Cliffs, NJ: McGraw-Hill.

27. A. N. Venkat, I. A. Hiskens, J. B. Rawlings, S. J. Wright. 2008. Distributed MPC strategies with application to power system automatic generation control. *IEEE Trans. Control Syst. Technol.* 16(6):1192–206.

28. H. Bevrani, T. Hiyama. 2009. On load-frequency regulation with time delays: Design and real-time implementation. *IEEE Trans. Energy Conversion* 24:292–300.

29. D. Heckerman. 1997. Bayesian networks for data mining. *Data Mining Knowledge Discovery* 1:79–119.

30. F. Daneshfar, H. Bevrani. 2010. Load-frequency control: A GA-based multi-agent reinforcement learning. *IET Gener. Transm. Distrib.* 4(1):13–26.

31. H. Bevrani, F. Daneshfar, P. R. Daneshmand. 2010. Intelligent power system frequency regulation concerning the integration of wind power units. In *Wind power systems: Applications of computational intelligence*, ed. L. F. Wang, C. Singh, A. Kusiak, 407–37. Springer Book Series on Green Energy and Technology. Heidelberg: Springer.

32. H. Bevrani, T. Hiyama, H. Bevrani. 2011. Robust PID based power system stabilizer: Design and real-time implementation. *Electrical Power and Energy Syst.* 33: 179–88.

9

Fuzzy Logic and AGC Systems

An overview on fuzzy-logic-based AGC systems with different configurations is given in Chapter 3. This chapter presents two multifunctional fuzzy-logic-based AGC schemes considering the MWh constraint for the power transmission through the interarea tie-line and the regulation margin for the AGC units. The first design scheme includes two control loops. The main control loop consists of a three-dimensional polar-information-based fuzzy logic control block, and the additional control loop consists of a two-dimensional polar-information-based fuzzy logic control block.[1,2] There exist additional parameters, which give the contribution factor from each subarea and another contribution factor from each unit. These contribution factors are determined according to the remaining available capacities of the AGC units in each subarea, and that of each individual AGC unit, respectively. Additional functions, such as the real power flow control scheme on the trunk lines considering the power flow constraints, and the regulation margin control scheme to maintain the remaining AGC capacity to the specified level,[3,4] have also been proposed. These additional control loops are activated only when the power flow constraints are violated, and also only when the level of the regulation margin becomes lower than the specified level.

The second design provides a particle-swarm-optimization (PSO)-based fuzzy logic AGC system. In order to improve the control performance, the PSO technique is used for tuning of the fuzzy system's membership function parameters in the supplementary frequency control loop.[5] The efficiency of the proposed control schemes has been demonstrated through nonlinear simulations by using appropriate interconnected power system examples.

9.1 Study Systems

9.1.1 Two Control Areas with Subareas

The study system for the first AGC synthesis approach is shown in Figure 9.1. The study system consists of two areas. Area X is the study area, and area Y is its external area. Study area X is divided into four subareas: A, B, C, and D. Each subarea has a certain number of AGC units for the load-frequency regulation and also non-AGC units. There are several trunk lines between these subareas; therefore, the power accommodation between the subareas is

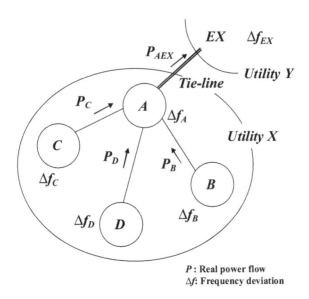

FIGURE 9.1
Interconnected two-area power system.

also possible through the trunk lines, in addition to the power accommodation between area X and area Y.

A detailed nonlinear AGC simulator has been developed in the MATLAB/Simulink environment.[6] The simulator includes twenty-three thermal units to simulate the AGC performance of the entire Kyushu Electric Power System in Japan. However, the power generation from the nuclear units, from the hydro units, and also from the other utility units in the Kyushu Electric Power System is not included in the simulation model because these units are out of the AGC.[7] In the simulations, the generation rate constraints are considered for each unit, and various types of load changes, such as step, ramp, and random changes, are utilized.

All the nonlinear simulations have been performed by using the detailed simulation program developed in the MATLAB/Simulink environment. Figure 9.2 illustrates the main Simulink block for the study system. There exist five subblocks that represent the external area EX, and the subareas A to D. Each subblock also consists of several blocks, such as a power generation block and block for the load-frequency dynamics. The main block also includes a power flow calculation block to determine the real power flow on the tie-line and on the trunk lines. This test system is used to examine the polar-information-based fuzzy logic AGC scheme.

9.1.2 Thirty-Nine-Bus Power System

To investigate the performance of the PSO-based fuzzy logic AGC design, a simulation study is provided in the SimPowerSystems environment of

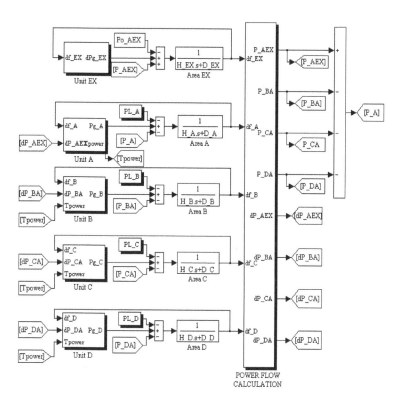

FIGURE 9.2
Main Simulink block.

MATLAB software. For this purpose, a network with the same topology as the well-known IEEE ten-generators, thirty-nine-bus test system, described in Figure 6.7, is considered. The thirty-nine-test system is organized into three areas. Here, to increase the power fluctuation, all three areas are equipped with wind farms. A single line diagram of the updated test system is redrawn in Figure 9.3. This system has ten generators, nineteen loads, thirty-four transmission lines, and twelve transformers. The simulation parameters for the generators, loads, lines, and transformers of the test system are given in Bevrani et al.[8] The installed wind farms use a double-fed induction generator (DFIG) wind turbine type.

The total installed wind power is about 85 MW of wind power generation. Similar to the simulation studies of Chapters 7 and 8, it is assumed that all power plants in the power system are equipped with a speed governor and power system stabilizer (PSS). However, only one generator in each area is responsible for the AGC issue: G1 in area 1, G9 in area 2, and G4 in area 3.

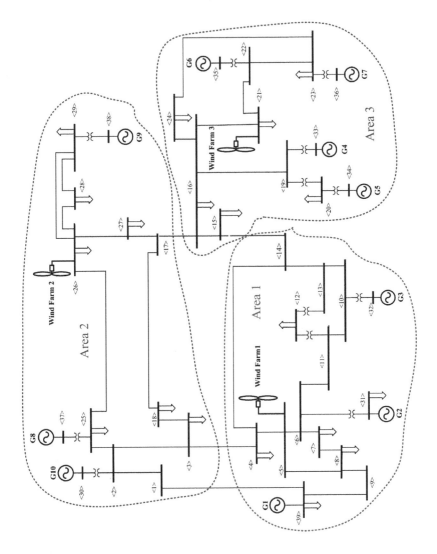

FIGURE 9.3
Single line diagram of updated thirty-nine-bus case study.

For the sake of simulation, random variations of wind velocity have been considered. The dynamics of wind turbine generators, including the pitch angle control of the blades, are also considered.

9.2 Polar-Information-Based Fuzzy Logic AGC

9.2.1 Polar-Information-Based Fuzzy Logic Control[1,2]

According to the frequency change, the tie-line power change, and the integrated tie-line power deviation, the total demand of the additional generation is determined through a simple polar-information-based fuzzy logic control scheme in the main control loop shown later. Three-dimensional information is required to determine the additional generation as follows:

$$Za(k) = (ACE(k) - ACE(k - 1))/\Delta T \tag{9.1}$$

$$Zs(k) = ACE(k) \tag{9.2}$$

$$Zp(k) = \Sigma \Delta P_{AEX}(k) \tag{9.3}$$

where

$$ACE(k) = \Delta P + \beta \Delta f \tag{9.4}$$

Here, the sampled variables $Zs(k)$, $Za(k)$, and $Zp(k)$ represent the area control error (ACE), the ACE gradient, and the integrated tie-line power deviation, respectively. The area control error $ACE(k)$ determines the frequency deviation and the total power flow change from each subarea through the tie-line and the trunk lines.

Figure 9.4 illustrates the switching surface utilized to determine the additional generation for the AGC. The state space is divided into two sub-spaces. In the space over the switching surface, the total generation should be reduced to keep the frequency and the tie-line power. On the contrary, the total generation should be increased in the space below the switching surface. In order to simplify the control algorithm using the three-dimensional information, the system state is replaced onto a two-dimensional phase plane, shown in Figure 9.5. Then, the system state is given by a point $p(k)$ in the phase plane.

$$p(k) = [Z_s(k) + ZSS, A_s Z_a(k)] \tag{9.5}$$

where $ZSS(k) = Sg \cdot Zp(k)$.

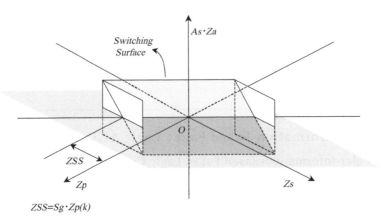

FIGURE 9.4
Switching surface for fuzzy logic control.

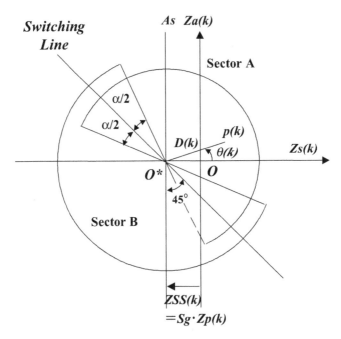

FIGURE 9.5
Phase plane to identify system state.

The term $ZSS(k)$ gives the shift size of the origin O to the origin O^* in the transient period, and Sg is the shift gain. At the final steady state, the origin O^* coincides with the origin O, because $Zp(k)$ becomes equal to zero at the final steady state. The polar notation of the present state is given by the following two equations:

$$D(k) = \sqrt{(Zs(k)+ZSS)^2 + (As \cdot Za(k))^2} \tag{9.6}$$

$$\theta(k) = \tan^{-1}(As \cdot Za(k) / (Zs(k)+ZSS)) \tag{9.7}$$

Finally, the defuzzification process to provide the control signal is described as follows:

$$u(k) = \frac{N(\theta(k)) - P(\theta(k))}{N(\theta(k)) + P(\theta(k))} \cdot G(D(k)) \cdot U_{max} \tag{9.8}$$

$$G(D(k)) = D(k)/Dr \quad for \quad D(k) \leq Dr \tag{9.9}$$

$$G(D(k)) = 1.0 \quad for \quad D(k) \geq Dr \tag{9.10}$$

where U_{max} is the maximum size of the control signal from the fuzzy logic control block.

Here, the integrated tie-line power deviation $Zp(k)$ is utilized to determine the additional generation demand to maintain the integrated tie-line power deviation at zero. In other words, the contract MWh constraint is satisfied through the proposed fuzzy logic control scheme. The final total power demand is given by *Tpower*, which is the integration of the control signal $u(k)$ in the main regulation block, shown in Figure 9.6.

In Figure 9.6, the term *pf_A* gives the participation factor of subarea *A* to the load-frequency regulation, and therefore the power generation in subarea *A* is changed to the value of *Tpower·pf_A*. The power generation in the other subareas is also changed according to the values of *Tpower·pf_B*, *Tpower·pf_C*, and *Tpower·pf_D*, as shown in Figure 9.7. Each participation factor is determined based on the available AGC capacity in the associated area. Figure 9.8 also includes an additional control loop to regulate the power flow on the trunk line from the associated subarea. Here, a fuzzy-logic-based control scheme is utilized, where only two-dimensional information, i.e., the ACE and its gradient for the associated area are used. The control signal is determined by using the same equations shown above after setting the shift gain *Sg* to zero.

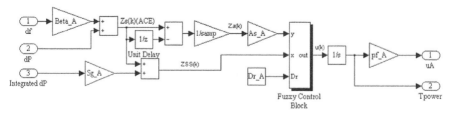

FIGURE 9.6
Main regulation block.

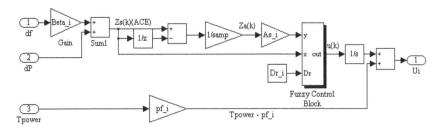

FIGURE 9.7
Subregulation block for subareas *B*, *C*, and *D*.

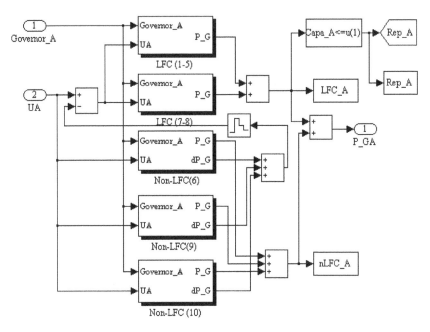

FIGURE 9.8
Power generation block, including AGC and non-AGC power stations.

The additional control loop is activated only when the real power flow constraint on the trunk line is violated. The configuration of the control loop is the same as that of the conventional supplementary control loop; therefore, the load change in the associated area is regulated by the generation change in the same area. Consequently, the trunk line power is maintained up to its power flow limit.

Figure 9.8 illustrates the power generation block, which includes several AGC and non-AGC power stations in each subarea. Each subarea has a specific number of AGC units and non-AGC units. Whenever the power generation from the AGC units exceeds the prespecified limit, a portion of the power generation is shifted to the non-AGC units in order to keep the regulation

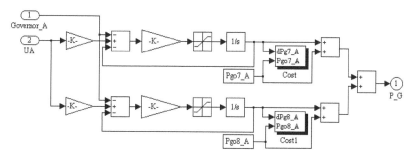

FIGURE 9.9
Configuration of an AGC power station.

margin to a certain level. Namely, the regulation capacity is always kept to a certain extent by the participation of the non-AGC unit to the load-frequency regulation. Usually, the generation from the non-AGC unit can be changed quite slowly; therefore, it takes some time to finish the replacement of the required generation from the AGC unit to the non-AGC unit. However, this does not mean that the regulation requires the same length of time to reach the steady state. This issue is shown later.

Figure 9.9 shows the detailed configuration of one of the AGC power stations, where there exist two AGC units. Each AGC unit uses a specific participation factor, but the total sum of participation factors among an area is equal to 1.0.

9.2.2 Simulation Results

Simulations have been performed subject to various types of load changes, such as step, ramp, and random load. All the control parameters have been set to their typical values; therefore, the tuning of these parameters is not considered in this study. Throughout the simulations, the frequency deviation, tie-line power, and trunk line power are sampled every 2.5 s, and the load-frequency regulation command is renewed at every 10 s, considering the practical AGC operation.

9.2.2.1 Trunk Line Power Control

Table 9.1 shows the initial settings of the total power generation from the AGC units, non-AGC units, and all units. In subarea *D*, there is no non-AGC unit. Table 9.2 shows the real power flow on the tie-line and also on the trunk lines. Table 9.3 indicates the load change considered for the simulations.

When the trunk line power control is not considered, the total load change in the utility *X* is covered by the generation changes from all the units, including the non-AGC units, to keep the regulation capacity to a certain extent. The contribution from each unit is determined by the subarea participation factors together with the unit participation factors in each subarea. In Figure 9.10, from the top to the bottom, the real power flows on the tie-line

TABLE 9.1

Initial Setting of Total Generation (1 pu = 10,000 MW)

Generation	Area A	Area B	Area C	Area D
From AGC units (pu)	0.32	0.08	0.40	0.20
From non-AGC units (pu)	0.06	0.02	0.10	—
Total generation (pu)	0.40	0.10	0.50	0.20

TABLE 9.2

Tie-Line Power and Trunk Line Power (1 pu = 10,000 MW)

Tie-Line	P_{AEX}	P_B	P_C	P_D
Trunk Lines	A-EX	B-A	C-A	D-A
Power flow (pu)	0.10	0.00	0.20	0.10

TABLE 9.3

Load Change

	Area A	Area B	Area C	Area D
Initial load (pu)	0.60	0.10	0.30	0.10
Type and size (pu)	Random	—	Ramp: –0.1	—
Start time (s)	Always	—	50	—

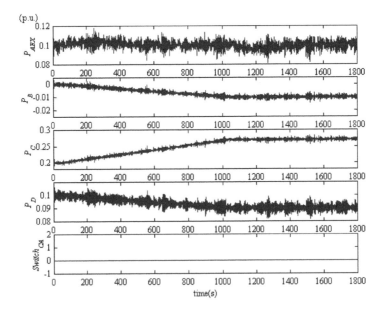

FIGURE 9.10
Real power flow on tie-line and trunk lines without trunk line power control.

and the trunk lines, from subareas *B*, *C*, and *D* to subarea *A*, are illustrated. In the bottom graph, the activation of the trunk line power control is indicated. The value 0 shows the nonactivation, while the value 1 shows the activation. In this simulation, the real power flow limit is not specified; therefore, the trunk line power control is not activated.

In order to check the efficiency of the proposed trunk line power control, the real power limit is specified at 0.22 pu from subarea *C* to the subarea *A*. The simulation results are shown in Figures 9.11 and 9.12. Figure 9.11 shows the real power flow on the tie-line and also on the trunk lines. As shown in the variation of the real power flows and in the bottom graph, the trunk line power control is activated whenever the trunk line power exceeds the specified limit of 0.22 pu, and in the final steady state the trunk line power is kept at its limit of 0.22 pu. Figure 9.12 shows the frequency deviation in each subarea. The frequency deviation is within the range of tolerance, i.e., 0.05 Hz. Furthermore, the profile of frequency change in each subarea has almost the same profile as shown in Figure 9.12.

9.2.2.2 Control of Regulation Margin

In the following simulations, the efficiency of the regulation margin control has been demonstrated. The load change utilized for the simulation is shown

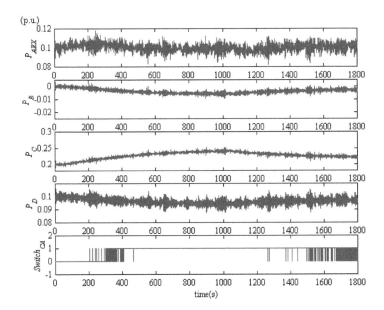

FIGURE 9.11
Real power flow on tie-line and trunk lines with trunk line power control.

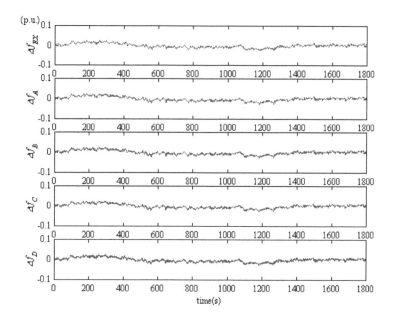

FIGURE 9.12
Frequency deviation.

TABLE 9.4

Load Change

	Area A	Area B	Area C	Area D
Initial load (pu)	0.60	0.10	0.30	0.10
Type and size (pu)	Random-ramp: 0.03 pu	—	—	—
Start time (s)	50 (s)	—	—	—

in Table 9.4. In the simulation, the power generation limits from the AGC units in subarea A are specified at 0.33 pu. Whenever the generation from the AGC units exceeds this limit, a certain power generation is shifted from the AGC units to the non-AGC units to keep the generation from the AGC units to its limit of 0.33 pu in the final steady state.

Figure 9.13 shows the variations of the total power generation from the AGC units in subareas A to D. The bottom graph illustrates the activation of the control of the regulation margin, where the value zero indicates the non-activation, while the value 1 indicates the activation. As shown in this figure, the power generation from the AGC units in subarea A is reduced by the proposed control scheme to have the specified regulation margin. Here, it must be noted that the total power generation from the AGC units exceeds the specified limit in the transient stage after having the load change. Figure 9.14

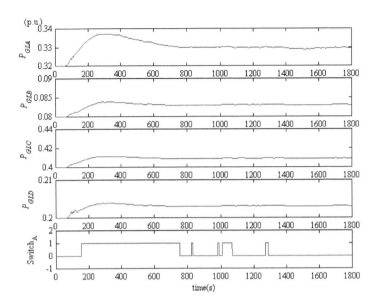

FIGURE 9.13
Total generation from AGC units in subareas *A* to *D*.

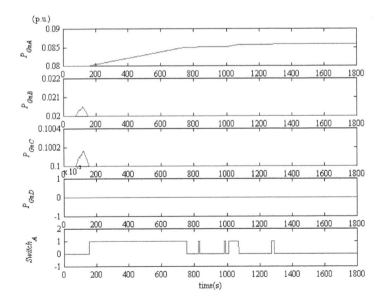

FIGURE 9.14
Total generation from non-AGC units in subareas *A* to *D*.

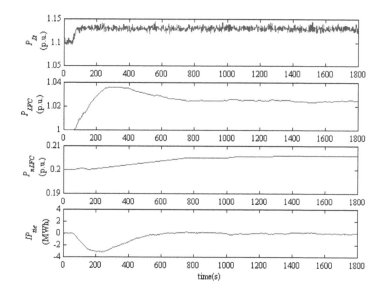

FIGURE 9.15
Total load, total generation from AGC units, total generation from non-AGC units, and integrated tie-line power deviation.

shows the total generation from the non-AGC units in subareas *A* to *D*. In the subareas, a certain amount of power comes from the non-AGC units in the final steady state to keep the regulation margin.

Figure 9.15 illustrates the total load, the total generation from the AGC units, the total generation from the non-AGC units, and the integrated tie-line power deviation from the top to the bottom. As shown in the figure, the integrated tie-line power deviation becomes equal to zero in the final steady state. Namely, the MWh constraint is satisfied by using the proposed control scheme.

9.3 PSO-Based Fuzzy Logic AGC

9.3.1 Particle Swarm Optimization

Particle swarm optimization (PSO) is a population-based stochastic optimization technique. It belongs to the class of direct search methods that can be used to find a solution to an optimization problem in a search space. The PSO was originally presented based on the social behavior of bird flocking, fish schooling, and swarming theory.[9,10] In the PSO method, a swarm consists of a set of individuals, with each individual specified by *position* and *velocity* vectors $(x_i(t), v_i(t))$ at each time or iteration. Each individual is named a *particle*, and the position of every particle represents a potential solution to the under-study optimization problem. In an n-dimensional solution space, each particle is treated as an n-*dimensional* space vector and the position of the ith particle is presented by $v_i = (x_{i1}, x_{i2}, ..., x_{in})$; then it flies to a new position by the velocity represented by $v_i = (v_{i1}, v_{i2}, ..., v_{in})$. The best position for ith particle represented by $p_{best,i} = (p_{best,i1}, p_{best,i2}, ..., p_{best,in})$ is determined according to the best value for the specified objective function.

Furthermore, the best position found by all particles in the population (global best position) can be represented as $g_{best} = (g_{best,1}, g_{best,2}, ..., g_{best,n})$. In each step, the best particle position, global position, and corresponding objective function values should be saved. For the next iteration, the position x_{ik} and velocity v_{ik} corresponding to the kth dimension of ith particle can be updated using the following equations:

$$v_{ik}(t+1) = w.v_{ik} + c_1.rand_{1,ik}(p_{best,ik}(t) - x_{ik}(t)) + c_2.rand_{2,ik}(g_{best,k}(t) - x_{ik}(t)) \quad (9.11)$$

$$x_{ik}(t+1) = x_{ik}(t) + v_{ik}(t+1) \quad (9.12)$$

where $i = 1, 2, ..., n$ is the index of particles, w is the inertia weight,[11] $rand_{1,ik}$ and $rand_{2,ik}$ are random numbers in the interval [0 1], c_1 and c_2 are learning factors, and t represents the iterations.

Usually, a standard PSO algorithm contains the following steps:

1. Initialize all particles via a random solution. In this step, each particle position and its associated velocity are set by randomly generated vectors. The dimension of the position should be generated within a specified interval, and the dimension of the velocity vector should also be generated from a bounded domain using uniform distributions.

2. Compute the objective function for the particles.

3. Compare the value of the objective function for the present position of each particle with the value of the objective function corresponding to the prespecified best position, and replace the prespecified best position with the present position, if it provides a better result.

4. Compare the value of the objective function for the present best position with the value of the objective function corresponding to the global best position, and replace the present best position by the global best position, if it provides a better result.

5. Update the position and velocity of each particle according to Equations 9.11 and 9.12.

6. Stop the algorithm if the stop criterion is satisfied. Otherwise, go to step 2.

In the present section, the PSO algorithm is used to find the optimal value for membership function parameters of a fuzzy-logic-based AGC system.

9.3.2 AGC Design Methodology

Inherent nonlinearity, increasing in size and complexity of power systems as well as emerging wind turbines and their effects on the dynamic behavior of power systems, caused conventional proportional-integral-derivative (PID) and proportional-integral (PI) controllers to be incapable of providing good dynamical performance over a wide range of operating conditions.[12] In this section, to track a desirable AGC performance in the presence of high-penetration wind power in a multiarea power system, a decentralized PSO-based fuzzy logic control design is proposed. Decreasing the frequency deviations due to fast changes in output power of wind turbines, and limiting tie-line power interchanges to an acceptable range, following disturbances, are the main goals of this effort. The proposed control framework is shown in Figure 9.16.

The inputs and output are brought into an acceptable range by multiplying in proper gains. In each control area, *ACE* and its derivative are considered input signals, and the provided control signal is used to change the set points of AGC participant generating units. The *mamdani* type inference system is applied, and as shown in Figure 9.17, symmetric seven-segment triangular membership functions are used for input (a) and output (b) variables. The membership functions are defined as zero (ZO), large negative (LN), medium negative (MN), small negative (SN), small positive (SP), medium positive (MP), and large positive (LP).

Fuzzy rule is the basis of fuzzy logic operation to map the input space to the output space. Here, a rule base including forty-nine fuzzy rules is

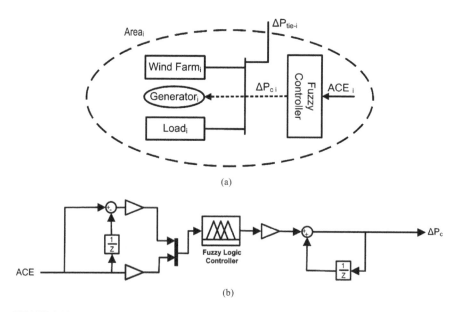

(a)

(b)

FIGURE 9.16
The proposed control framework: (a) area components and (b) controller structure.

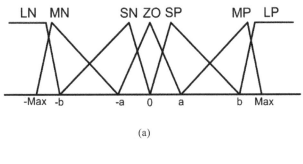

(a)

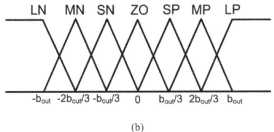

(b)

FIGURE 9.17
Symmetric fuzzy membership functions: (a) inputs pattern and (b) output pattern.

TABLE 9.5

Fuzzy Rule Base

		ΔACE						
		LN	**MN**	**SN**	**ZO**	**SP**	**MP**	**LP**
	LN	LP	LP	LP	MP	MP	SP	ZO
ACE	NM	LP	MP	MP	MP	SP	ZO	SN
	SN	LP	MP	SP	SP	ZO	SN	MN
	ZO	MP	MP	SP	ZO	SN	MN	MN
	S P	MP	SP	ZO	SN	SN	MN	LN
	MP	SP	ZO	SN	MN	MN	MN	LN
	LP	ZO	SN	MN	MN	LN	LN	LN

considered (Table 9.5). The rule base works on vectors composed of *ACE* and its gradient Δ*ACE*. Using Table 9.5, fuzzy rules can be expressed in the form of *if-then* statements, such as:

If ACE is SN and Δ*ACE* is MP, *then* output is SN.

As explained in Chapter 3, since fuzzy rules are stated in terms of linguistic variables, crisp inputs should also be mapped to linguistic values using fuzzification. As the number of inputs is more than one (two inputs), fuzzified inputs in the *if* part of the rules must be combined to a single number. Here, the combination is done based on interpreting the *and* operator as a product of the membership values, corresponding to measured inputs. The obtained single number from the *if* part of each rule is used to compute the consequence of the same rule (implication), which is a fuzzy set. In this work, the *product* method has considered the implication method. In order to combine rules and make a decision based on all rules, the *sum* method is used. Finally, for converting an output fuzzy set of the fuzzy system to a crisp value, the *centroid* method is used for defuzzication.[13]

9.3.3 PSO Algorithm for Setting of Membership Functions

Like the performance of a fuzzy system influenced by membership functions, in order to achieve good performance by the controller, a PSO algorithm is established to find the optimal value for membership function parameters and the exact tuning of them. As can be seen in Figure 9.17, each set of input membership functions can be specified by parameters a and b, where $min < a < b < max$. Also, for control output, one parameter needs to be specified. Therefore, five parameters should be optimized for input membership functions using the PSO algorithm: $a_{in,\ ACE}$, $b_{in,\ \Delta ACE}$, $a_{in,\ \Delta ACE}$, $b_{in,\Delta ACE}$, and b_{out}.

For the sake of the PSO algorithm in the present AGC design, the objective function (f) is considered as given in Equation 9.13. The number of particles, particle size, and v_{min}, v_{max}, c_1, and c_2 are chosen as 10, 6, −0.5, 0.5, 2.8, and 1.3, respectively. Following use of the PSO algorithm, the optimal values for membership function parameters are obtained as listed in Table 9.6.

$$f = \frac{1}{3}\sum_{i=1}^{3}\left(\int t\left(|\Delta f_i| + |\Delta P_{tie,i}|\right)dt\right)$$

(9.13)

9.3.4 Application Results

To demonstrate the effectiveness of the proposed fuzzy-logic-based AGC design, some simulations were carried out. In these simulations, the proposed controllers were applied to the model described in Section 9.1.2, and

TABLE 9.6

Optimal Values of Membership Function Parameters

$a_{in,\ ACE}$	$b_{in,\ ACE}$	$a_{in,\ \Delta ACE}$	$b_{in,\ \Delta ACE}$	b_{out}
0.267747	0.947038	0.013716	0.059880	0.986659

also used in the literature.[5,8,14,15] In the performed simulation, the performance of the closed-loop system uses the well-tuned conventional PI controllers, compared to the designed fuzzy-logic-based controllers. As a serious test scenario, the following load disturbances (step increase in demand) are applied to three areas at 5 s simultaneously: 3.8% of the total area load at bus 9 in area 1, 4.3% of the total area load at bus 18 in area 2, and 8.01% of the total area load at bus 24 in area 3. The simulation is implemented by using the MATLAB SimPowerSystems program. The simulation results are shown in Figures 9.18 to 9.21.

The total generated wind power, wind speed deviation, and overall frequency deviation of the system are shown in Figure 9.18. The area control error signals of three areas, following the load disturbances, are also shown in Figure 9.19. The produced mechanical power of AGC participant generating units and the tie-line powers are illustrated in Figures 9.20 and 9.21, respectively.

These figures show the superior performance of the proposed fuzzy logic method to the conventional PI controller in deriving area control error and frequency deviation close to zero. Interested readers can find more details on the proposed PSO-based fuzzy logic AGC design method with numerous simulation results for the same case study in Daneshmand.[5]

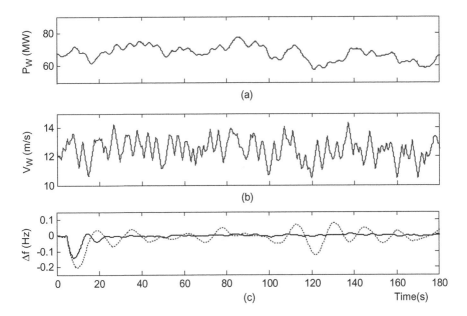

FIGURE 9.18
System response: (a) total wind power generation, (b) wind speed deviation, and (c) frequency deviation (solid, proposed fuzzy control; dotted, conventional PI control).

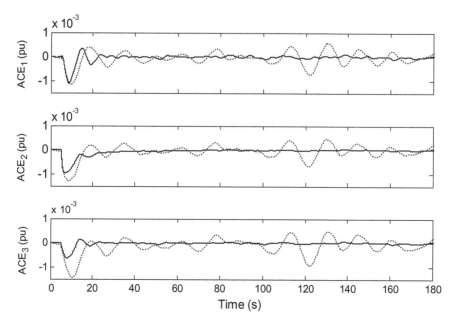

FIGURE 9.19
ACE signal in three control areas, following simultaneous disturbances; solid (proposed fuzzy control), dotted (conventional PI control).

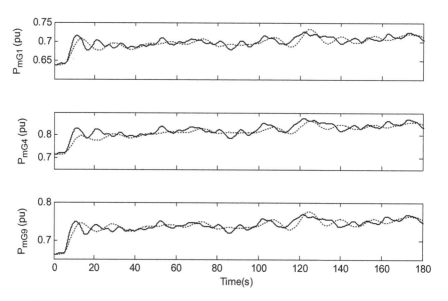

FIGURE 9.20
Mechanical power of AGC participating units, following simultaneous disturbances; solid (proposed fuzzy control), dotted (conventional PI control).

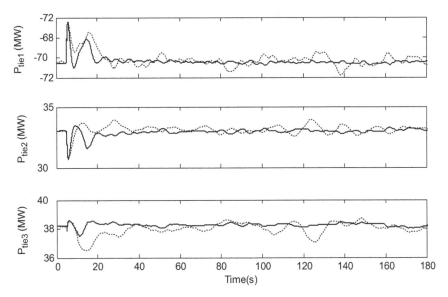

FIGURE 9.21
Tie-line powers, following simultaneous disturbances; solid (proposed fuzzy control), dotted (conventional PI control).

9.4 Summary

Two fuzzy-logic-based AGC design methodologies have been presented for the frequency and tie-line power regulation in multiarea power systems. These methodologies are polar-information-based fuzzy logic AGC and PSO-based fuzzy logic AGC designs. The efficiency of the proposed control schemes has been demonstrated through nonlinear simulations by using the detailed models developed in the MATLAB/Simulink and SimPowerSystems environments.

By using the proposed polar-information-based fuzzy logic AGC scheme, the MWh constraint is satisfied to avoid the MWh contract violation. The regulation capacity is always kept to a certain level by the replacement of the required generation from the AGC units to the non-AGC units. In addition, the power flow on the trunk lines can be regulated whenever the power flow exceeds the specified limit. The PSO-based fuzzy logic AGC design is used for frequency and tie-line power regulation in the presence of wind turbines. The proposed method is applied to an updated version of the ten-generator, thirty-nine-bus system, and the results are compared with those from a conventional PI control design.

References

1. T. Hiyama, S. Oniki, H. Nagashima. 1996. Evaluation of advanced fuzzy logic PSS on analog network simulator and actual installation on hydro generators. *IEEE Trans. Energy Conversion* 11(1):125–31.
2. T. Hiyama, T. Kita, T. Miyake, H. Andou. 1999. Experimental studies of three dimensional fuzzy logic power system stabilizer on damping of low-frequency global mode of oscillation. *Fuzzy Sets Systems* 102(1):103–11.
3. T. Hiyama. 1982. Optimisation of discrete-type load-frequency regulators considering generation rate constraints. *IEE Proc. C* 129(6):285–89.
4. T. Hiyama. 1981. Design of decentralised load-frequency regulators for interconnected power systems. *IEE Proc. C* 129(1):17–23.
5. P. R. Daneshmand. 2010. Power system frequency control in the presence of wind turbines. MSc dissertation, Department of Electrical and Computer Engineering, University of Kurdistan, Sanandaj, Iran.
6. T. Hiyama, Y. Yoshimuta. 1999. Load-frequency control with MWh constraint and regulation margin. In *Proceedings of IEEE Power Engineering Society, 1999 Winter Meeting, New York*, vol. 2, pp. 803–8.
7. H. Bevrani, G. Ledwich, Z. Y. Dong, J. J. Ford. 2009. Regional frequency response analysis under normal and emergency conditions. *Elect. Power Syst. Res.* 79:837–45.
8. H. Bevrani, F. Daneshfar, P. R. Daneshmand. 2010. Intelligent power system frequency regulation concerning the integration of wind power units. In *Wind power systems: Applications of computational intelligence*, ed. L. F. Wang, C. Singh, A. Kusiak, 407–37. Springer Book Series on Green Energy and Technology. Heidelberg: Springer-Verlag.
9. J. Kennedy, R. Eberhart. 1995. Particle swarm optimization. In *Proceedings of IEEE International Conference, Neural Networks*, vol. 4, pp. 1942–48.
10. R. Eberhart, J. Kennedy. 1995. A new optimizer using particle swarm theory. In *Proceedings of Sixth International Symposium, Micro Machine and Human Science, Nagoya, Japan*, pp. 39–43.
11. Y. Shi, R. Eberhart. 1998. A modified particle swarm optimizer. In *Proceedings of IEEE International Conference, Evolutionary Computation, IEEE World Congress, Computational Intelligence*, Anchorage, AK, May 4–9, 1998, pp. 69–73.
12. H. Bevrani. 2009. *Robust power system frequency control.* New York: Springer.
13. H. Bevrani, Syafaruddin, T. Hiyama. Intelligent control in power systems. Lecture notes. http://www.cs.kumamoto-u.ac.jp/epslab/sub1.html.
14. H. Bevrani, F. Daneshfar, P. R. Daneshmand, T. Hiyama. 2010. Reinforcement learning based multi-agent LFC design concerning the integration of wind farms. In *Proceedings of IEEE International Conference on Control Applications, Yokohama, Japan,* CD-ROM.
15. H. Bevrani, F. Daneshfar, P. R. Daneshmand, T. Hiyama. 2010. Intelligent automatic generation control: Multi-agent Bayesian networks approach. In *Proceedings of IEEE International Conference on Control Applications, Yokohama, Japan,* CD-ROM.

10

Frequency Regulation Using Energy Capacitor System

As the use of *energy capacitor systems* (ECSs) and other energy storage devices increases worldwide, there is a rising interest in their application in power system operation and control, especially in power system frequency regulation.[1-6] Frequency regulation as an ancillary service associated with AGC systems is becoming increasingly important for supplying sufficient and reliable electric power, in keeping high power quality under the deregulated power systems, including independent power producers.[7,8] As mentioned in Chapter 2, to ensure satisfactory area frequency, interarea tie-line power is one of the main requirements for frequency regulation. The tie-line bias control (TBC) through the supplementary feedback loop plays a significant role to meet the mentioned requirement.[9,10]

This chapter presents a coordinated frequency regulation scheme between the conventional AGC participating units and a small-sized ECS to improve the frequency regulation performance. The ECS consists of electrical double-layer capacitors. One of its specific features is the fast charging/discharging operation with a high power level. In the proposed scheme, the charging/discharging operation is performed on the ECS to balance the total power generation with the total power demand. The ECS can absorb the rapid variation of load power for balancing the generation with the demand. However, a small-sized ECS is assumed in this study; therefore, the continuous discharging and charging operations are not available on the ECS because the stored energy level hits its lower or upper limit. To perform the continuous frequency regulation on the small-sized ECS, coordination by the conventional AGC units has been proposed to keep the stored energy level on the ECS within the prespecified level. The power generation from the AGC units increases whenever the stored energy level on the ECS decreases. On the contrary, the power generation from the AGC unit decreases whenever the stored energy level on the ECS increases. By the proposed coordination, the frequency regulation performance is highly improved. To demonstrate the efficiency of the proposed control scheme, nonlinear detailed simulations have been performed for a two-area interconnected system with a practical size in the MATLAB/Simulink environment. The simulation results clearly indicate the advantages of the proposed control scheme.

10.1 Fundamentals of the Proposed Control Scheme

According to the frequency and tie-line power changes, the total demand of the additional generation is determined. To determine the additional generation, the area control error, ΔACE, as a linear combination of frequency deviation, Δf, and tie-line power change, ΔP_{tie}, is defined (Equation 2.11).

In this study, an innovative frequency regulation scheme has been proposed considering the coordination between the ECS and the conventional AGC units. The basic configuration of the proposed controller for the ECS is shown in Figure 10.1. The charging/discharging level, U_{ecs}, is specified by using the ΔACE monitored in the study area, for an output setting of ECS.

From the monitored area control error, the charging/discharging level on the ECS is determined to balance the total generation with the total demand. Following the control signal U_{ecs} from the *PI* control loop, the charging/discharging operation is performed on the ECS for the frequency regulation. Because of the specific feature of the ECS dynamics, the fast charging/discharging operation is available on the ECS. Therefore, even the rapid variations of demand can be efficiently absorbed through the charging or discharging operation on the ECS. Since small-sized ECS is considered in this study, the regulation power from the conventional AGC units is required to keep the stored energy level of the ECS unit in a proper range and not to prevent the AGC action on the ECS.

Figure 10.2 illustrates the configuration of the coordinated controller for the conventional AGC units. In the proposed control scheme, the ECS provides

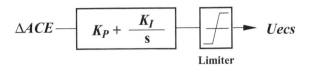

FIGURE 10.1
Control block for the proposed regulation scheme on the ECS unit.

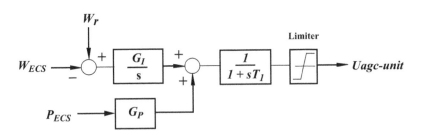

FIGURE 10.2
Coordinated control structure for the AGC unit.

its main function and the AGC units provide its supplementary function, not to prevent the charging/discharging operation on the ECS. The coordinated frequency regulation between the ECS and AGC units has been considered to balance the power demand and total power generation. Namely, the power generation from the AGC units is regulated to maintain the stored energy level on the ECS for the uninterruptible frequency regulation on the small-sized ECS. In Figure 10.2, W_r and W_{ECS} are defined as target stored energy and current stored energy, respectively. The P_{ECS} is the produced power by ECS, and $U_{agc-unit}$ represents the required regulation power from the AGC unit for coordination with ECS. The power regulation command $U_{agc-unit}$ is utilized for the power regulation from the AGC units.

10.1.1 Restriction of Control Action (Type I)

The charging/discharging operation is performed by the control signal $Uecs$ determined through the *PI* control block as shown in Figure 10.1. Whenever the stored energy reaches its upper or lower limit, the tracking of the load variation will stop until the stored energy level comes back to its prespecified operation range and the charging/discharging operation starts again on the ECS. Namely, there exist some impacts to the study system when the stored energy level hits its upper or lower limit.

To overcome this situation, restriction of the control signal $Uecs$ has been proposed. Figure 10.3 illustrates one of the restrictions as type I to prevent the impact when the stored energy reaches its limits. The detailed restriction shown in Figure 10.3 is mathematically expressed as depicted in Table 10.1. The upper or lower limit of the control signal $Uecs$ is modified when the stored energy level reaches its limit to prevent the unnecessary control action. In this figure, [$-dWess\_max$ $+dWess\_max$] represents the range of acceptable operation.

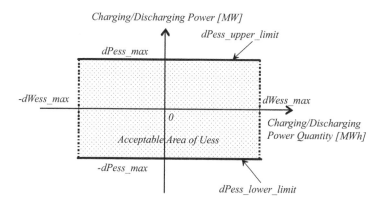

FIGURE 10.3
Restriction of the control signal to prevent excessive control action (type I).

TABLE 10.1

Restriction of Control Signal (Type I)

Condition	Restriction
$\Delta Wecs\_max \le \Delta Wecs$	$Uecs\_upper\_limit = Uecs\_max$
	$Uecs\_lower\_limit = 0$
$-\Delta Wecs\_max < \Delta Wecs < \Delta Wecs\_max$	$Uecs\_upper\_limit = Uecs\_max$
	$Uecs\_lower\_limit = -Uecs\_max$
$\Delta Wecs \le -\Delta Wecs\_max$	$Uecs\_upper\_limit = 0$
	$Uecs\_lower\_limit = -Uecs\_max$

10.1.2 Restriction of Control Action (Type II)

In the aforementioned type of restriction (type I), there still exist some impacts when the stored energy level reaches its upper or lower limit, because the control signal *Uecs* is set to zero from some existing nonzero signals in such conditions. Therefore, an alternative restriction has also been proposed for the further improvement to prevent unnecessary control actions when the stored energy level approaches its limits. Then, the upper or lower limit of the control signal *Uecs* is gradually modified to zero without giving any significant impact to the study system when the stored energy level reaches its upper or lower limit. The graphical description of this concept is shown in Figure 10.4. Table 10.2 gives the mathematical expression of the type II restriction for the upper or lower limit of the control signal *Uecs*. Here, *SP* shows the switching point. Y1 and Y2 in Table 10.2 are defined as follows:

$$Y1 = \frac{Uecs\_max}{\Delta Wecs\_max - SP}(\Delta Wecs - \Delta Wecs\_max) \tag{10.1}$$

$$Y2 = \frac{Uecs\_max}{\Delta Wecs\_max - SP}(\Delta Wecs + \Delta Wecs\_max) \tag{10.2}$$

10.1.3 Prevention of Excessive Control Action (Type III)

When the control action is restricted by applying a type I or type II restriction, there still exist some impacts caused by the accumulation of the area control error through the integral control loop shown in Figure 10.1. To overcome this situation, the control area itself should be modified to ΔACE^* according to the size of the control signal $Uecs^*$ after having type I or type II restriction, as shown in Figure 10.5. The mathematical expression for this restriction is shown in Table 10.3.

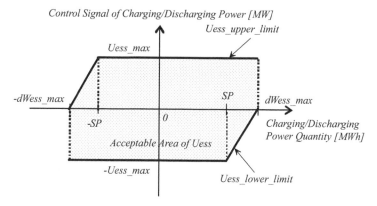

FIGURE 10.4
Restriction of the control signal to prevent excessive control action (type II).

TABLE 10.2

Restriction of Control Signal (Type II)

Condition	Restriction
$SP \le \Delta Wecs$	$Uecs\_upper\_limit = Uecs\_max$
	$Uecs\_lower\_limit = Y1$
$-SP < \Delta Wecs < SP$	$Uecs\_upper\_limit = Uecs\_max$
	$Uecs\_lower\_limit = -Uecs\_max$
$\Delta Wecs \le -SP$	$Uecs\_upper\_limit = Y2$
	$Uecs\_lower\_limit = -Uecs\_max$

10.2 Study System

Here, a two-area interconnected system, shown in Figure 9.1, is used as a study system. Utility X and utility Y are interconnected through the 500 kV tie-line. Utility X consists of four subareas, A to D. Subareas A, B, C, and D have eight, five, seven, and three thermal units, respectively. In addition, four and two nuclear units are operating in subareas C and D, respectively. Four hydro units are also included in subarea D. Five and three AGC units are in subareas A and B, respectively. The ECS is set in area A of utility X. Some nonlinear detailed simulations have been performed to demonstrate the efficiency of the proposed control scheme in the MATLAB/Simulink environment.

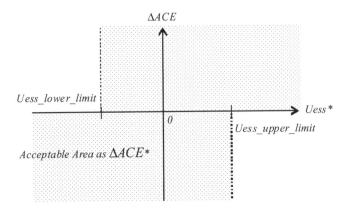

FIGURE 10.5
Restriction of the area control error to prevent excessive control action (type III).

TABLE 10.3

Restriction of Area Control Error (Type III)

Condition	Restriction
$Uecs\_upper\_limit \leq Uecs^*$	$\Delta ACE^* = (\Delta ACE \geq 0)$
$Uecs\_lower\_limit < Uecs^* < Uecs\_upper\_limit$	$\Delta ACE^* = \Delta ACE$
$Uecs^* \leq Uecs\_lower\_limit$	$\Delta ACE^* = (\Delta ACE \leq 0)$

10.3 Simulation Results

To evaluate the efficiency of the proposed coordinated AGC-ECS frequency regulation scheme, several nonlinear simulations have been performed following a random load change in subarea A and a ramp load change of 200 MW with a duration of 100 s. The capacity of the ECS is set to 1.6 MWh, and the maximum charging/discharging power is fixed at 120 MW.

Figure 10.6 illustrates the conventional frequency regulation performance. In the figure, the load change in subarea A, the load change in subarea B, the frequency deviation in subarea A, the deviation of the tie-line power, the power from the ECS, the deviation of the stored energy on the ECS, and the power change of the AGC (load-frequency control (LFC)) units in subareas A and B are illustrated from the top to the bottom. A relatively large fluctuation of the tie-line power of nearly 200 MW is caused by the random load change in subarea A.

Figure 10.7 shows the proposed frequency regulation performance with the control constraints of type I and type III. Figure 10.8 shows the proposed control system performance with the control constraints of type II and type III.

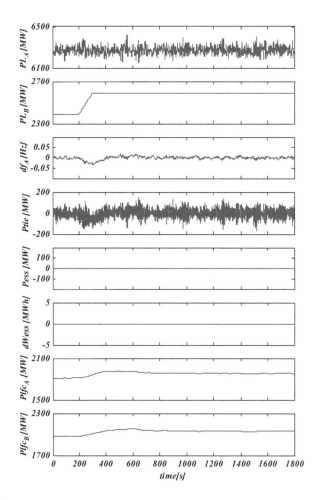

FIGURE 10.6
Performance of the conventional frequency regulation system.

The proposed control scheme is highly improved, especially when considering the type II and type III restrictions, as shown in Figure 10.8. Namely, the variations of the frequency deviation and the tie-line power deviation are small compared with those shown in Figure 10.6, where the conventional AGC is applied to the study system.

As shown in Figures 10.7 and 10.8, the stored energy level can be kept within the prespecified range through the coordination of the conventional AGC units. Therefore, a continuous AGC is available on the small-sized ECS. Furthermore, the AGC action is rocked for quite a short time range considering the restrictions of type II and type III to prevent the excessive control action, as shown in Figure 10.9.

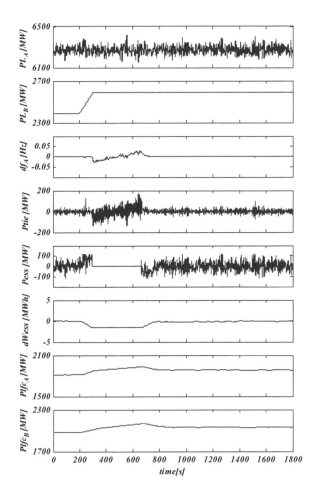

FIGURE 10.7
Performance of the proposed frequency regulation system (type I + type III).

10.4 Evaluation of Frequency Regulation Performance

In order to evaluate the AGC performance, the following performance index, J, is defined by using the area control error of area A.

$$J(\Delta ACE) = \Sigma \, \Delta ACE_A^2 \qquad (10.3)$$

The above performance index is evaluated under various situations on the capacity of the ECS and also on the maximum charging/discharging power to or from the ECS. The calculated performance index is normalized based on the index value obtained by the conventional frequency regulation

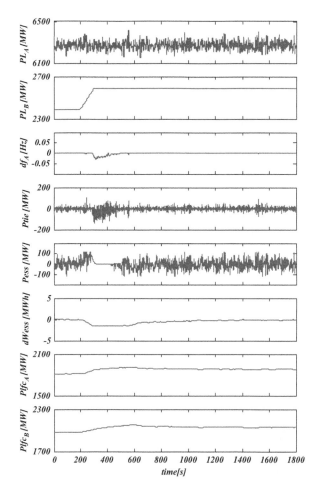

FIGURE 10.8
Performance of the proposed frequency regulation system (type II + type III).

scheme. Figure 10.9 illustrates the profile of the reduction of the performance index by applying the proposed frequency control scheme, where type I and type III restrictions are considered.

In Figure 10.9, the region of 0 to 20 indicates that the performance index is less than 20% of the conventional case. The region of 20 to 40 indicates that the performance index is greater than 20% and less than 40% of the conventional case, and so on. When considering type II and type III restrictions, the evaluated performance index is shown in Figure 10.10. The aspect of improvement is quite consistent according to the ECS capacity and the maximum power of the ECS. Namely, the improvement of the frequency regulation performance is further increasing for the larger ECS capacity and the larger power output of the ECS. From comparison of Figures 10.9 and 10.10, it is clear that the

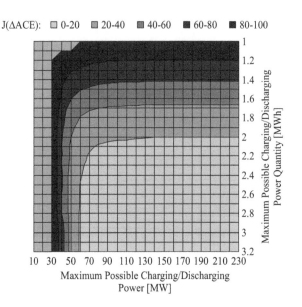

FIGURE 10.9
Evaluation of the frequency regulation performance (type I + type III).

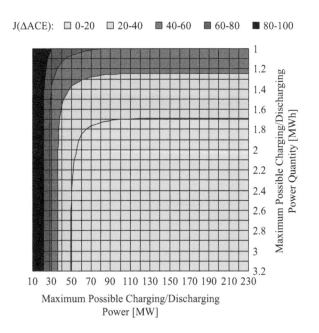

FIGURE 10.10
Evaluation of the frequency regulation performance (type II + type III).

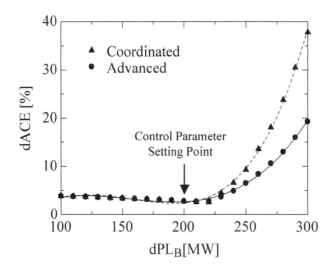

FIGURE 10.11
Effect of size of the load change on frequency regulation performance.

proposed control scheme with type II and type III restrictions is superior to achieve the better frequency regulation performance even on the small-sized and low-power ECS.

Figure 10.11 shows the variation of the performance index after changing the size of the ramp load in subarea B, where the capacity of the ECS is 1.6 MWh and the maximum power of the ECS is 120 MW. Here, it must be noted that all the control parameters have been determined for the ramp load change of 200 MW in subarea B. The proposed control scheme is highly robust when the restrictions of type II and type III are incorporated into it. The percentage of the performance index is less than 5% for the step load change, up to 225 MW.

10.5 Summary

A coordinated frequency regulation has been proposed for the small-sized and high-power energy capacitor system and the conventional AGC units. To prevent unnecessary excessive control action, two types of restrictions have been proposed for the upper and lower limits of the control signal as well as for the area control error. The simulation results clearly demonstrate the advantages of the proposed frequency regulation scheme. The control performance is highly improved through the proposed frequency regulation scheme.

References

1. M. Mufti, S. A. Lone, S. J. Iqbal, M. Ahmad, M. Ismail. 2009. Super-capacitor based energy storage system for improved load frequency control. *Elect. Power Syst. Res.* 79:226–33.
2. T. Sasaki, T. Kadoya, K. Enomoto. 2004. Study on load frequency control using redox flow batteries. *IEEE Trans. Power Syst.* 19(1):660–67.
3. S. C. Tripathy. 1997. Improved load-frequency control with capacitive energy storage. *Energy Conversion Manage.* 38(6):551–62.
4. S. K. Aditya, D. Das. 2001. Battery energy storage for load frequency control of an interconnected power system. *Elect. Power Syst. Res.* 58:179–85.
5. T. Hiyama, D. Zuo, T. Funabashi. 2002. Automatic generation control of stand alone power system with energy capacitor system. In *Proceedings of 5th International Conference on Power System Management and Control*, London, pp. 59–64.
6. M. Saleh, H. Bevrani. 2010. Frequency regulation support by variable-speed wind turbines and SMES. *World Acad. Sci. Eng. Technol.* 65:183–87.
7. I. Kumar, K. Ng, G. Shehle. 1997. AGC simulator for price-based operation. Part 1. A model. *IEEE Trans. Power Syst.* 12(2):527–32.
8. I. Kumar, K. Ng, G. Shehle. 1997. AGC simulator for price-based operation. Part 2. Case study results. *IEEE Trans. Power Syst.* 12(2):533–38.
9. T. Hiyama. 1982. Optomisation of discrete-type load frequency regulation considering generation rate constraints. *IEE Proc. C* 129(6):285–89.
10. T. Hiyama. 1982. Design of decentralised load-frequency regulation for interconnected power systems. *IEE Proc. C* 129(1):17–23.

11

Application of Genetic Algorithm in AGC Synthesis

Genetic algorithm (GA) is a numerical optimization algorithm that is capable of being applied to a wide range of optimization problems, guaranteeing the survival of the fittest. Time consumption methods such as trial and error for finding the optimum solution cause interest in meta-heuristic methods such as GA. The GA becomes a very useful tool for tuning of control parameters in AGC systems.

The GA begins with a set of initial random populations represented in chromosomes; each one consists of some genes (binary bits). These binary bits are suitably decoded to provide a proper string for the optimization problem. Genetic operators act on this initial population and regenerate the new populations to converge at the fittest. A function called *fitness function* is employed to aid regeneration of the new population from the older one, i.e., initial population. The fitness function assigns a value to each chromosome (solution candidate), which specifies its fitness. According to the fitness values, the results are sorted and some suitable chromosomes are employed to generate the new population by the specified operators. This process will continue until it yields the most suitable population as the optimal solution for the given optimization problem.

Several investigations have been reported in the past, pertaining to the application of GA in the AGC design.[1–15] Application of GA in AGC synthesis as a performance optimization problem in power system control is briefly reviewed in Chapter 3. In this chapter, following an introduction on the GA mechanism in Section 11.1, the GA application for optimal tuning of supplementary frequency controllers is given in Section 11.2. In Section 11.3, AGC design is formulated as a multiobjective GA optimization problem. A GA-based AGC synthesis to achieve the same robust performance indices as provided by the standard mixed H_2/H_∞ control theory is addressed in Section 11.4. The capability of GA to improve the learning performance in the AGC systems using a learning algorithm is emphasized in Section 11.5, and finally, the chapter is summarized in Section 11.6.

11.1 Genetic Algorithm: An Overview

11.1.1 GA Mechanism

The GA mechanism is inspired by the mechanism of natural selection, where stronger individuals would likely be the winners in a competing environment. Normally in a GA, the parameters to be optimized are represented in a binary string. A simplified flowchart for GA is shown in Figure 11.1. The *cost function*, which determines the optimization problem, represents the main link between the problem at hand (system) and GA, and also provides the fundamental source to provide the mechanism for evaluation of algorithm steps. To start the optimization, GA uses randomly produced initial solutions created by a random number generator. This method is preferred when a priori information about the problem is not available. There are basically three genetic operators used to produce a new generation: *selection, crossover,* and *mutation.* The GA employs these operators to converge at the global optimum. After randomly generating the initial population (as random solutions), the GA uses the genetic operators to achieve a new set of solutions at each iteration. In the selection operation, each solution of the current population is evaluated by its fitness, normally represented by the value of some objective function, and individuals with higher fitness values are selected.

Different selection methods such as stochastic selection or ranking-based selection can be used. In the selection procedure the individual chromosomes are selected from the population for the later recombination/crossover. The fitness values are normalized by dividing each one by the sum of all fitness

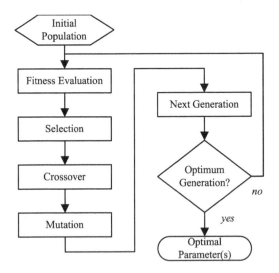

FIGURE 11.1
A simplified GA flowchart.

values named selection probability. The chromosomes with a higher selection probability have a higher chance to be selected for later breeding.

The crossover operator works on pairs of selected solutions with a certain crossover rate. The crossover rate is defined as the probability of applying a crossover to a pair of selected solutions (chromosomes). There are many ways to define the crossover operator. The most common way is called the *one-point crossover*. In this method, a point (e.g., for two binary-coded solutions of a certain bit length) is determined randomly in two strings and corresponding bits are swapped to generate two new solutions. A mutation is a random alteration with a small probability of the binary value of a string position, and it will prevent the GA from being trapped in a local minimum. The coefficients assigned to the crossover and mutation specify the number of the children. Information generated by the fitness evaluation unit about the quality of different solutions is used by the selection operation in the GA. The algorithm is repeated until a predefined number of generations have been produced.

Unlike the gradient-based optimization methods, GAs operate simultaneously on an entire population of potential solutions (chromosomes or individuals) instead of producing successive iterations of a single element, and the computation of the gradient of the cost functional is not necessary. Interested readers can find basic concepts and a detailed GA mechanism in Goldberg[16] and Davis.[17]

11.1.2 GA in Control Systems

GA is one of rapidly emerging optimization approaches in the field of control engineering and control system design.[18] Optimal/adaptive tracking control, active noise control, multiobjective control, robust tuning of control systems via seeking the optimal performance indices provided by robust control theorems, and use in fuzzy-logic- and neural-network-based control systems are some important applications of GA in control systems. Genetic programming can be used as an automated invention machine to synthesize designs for complex structures. It facilitates the design of robust dynamic systems with respect to environmental noise, variation in design parameters, and structural failures in the system.[19,20]

A simple GA-based control system is conceptually shown in Figure 11.2. The GA controller consists of three components: performance evaluator, learning algorithm, and control action producer. The performance evaluator rates a chromosome by assigning it a fitness value. The value indicates how good the chromosome is in controlling the dynamical plant to follow a reference signal. The learning algorithm may use a set of rules in the form of "condition then action" for controlling the plant. The desirable action will be performed by a control action producer when the condition is satisfied. The control structure shown in Figure 11.2 is implemented for several control applications in different forms.[21]

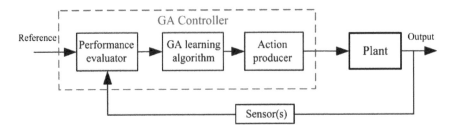

FIGURE 11.2
A GA-based control system.

11.2 Optimal Tuning of Conventional Controllers

Here, to show the capability of GA for tuning of a conventional integral controller in the supplementary frequency control loop, a simple three-area control system with a single thermal generator (reheat steam unit) and an integral controller in each control area is considered. The dynamic model of a governor-turbine, $M_i(s)$, and nominal parameters of systems are taken from Golpira and Bevrani[14] and Bevrani.[22]

The dynamic frequency response model for each area is considered as shown in Figure 11.3. The components of the block diagram are defined in Chapter 2. Here, the most important physical constraints are considered. For the sake of simulation, the generation rate constraint (GRC), governor deadband, and time delay in each supplementary frequency control loop are fixed at 3% pu.MW/min, 2 s, and 0.36 Hz, respectively.

For the present example, the initial population consists of one hundred chromosomes; each one contains forty-eight binary bits (genes). The fitness proportionate selection method (known as the roulette-wheel selection method) is used to select useful solutions for recombination. The crossover and mutation coefficients are fixed at 0.8 and 0.2. The objective function, which should be minimized, is considered as given in Equation 11.1:

$$J = \int_0^T (ACE_i)^2 \qquad (11.1)$$

where T is simulation time and $ACE_i = \Delta P_{tie,i} + \beta_i \Delta f_i$ is the area control error signal (see Equation 2.11). The applied GA steps are summarized as follows:

1. The initial population of one hundred random binary strings of length 48 has been built (each controller gains by sixteen genes).
2. The strings are decoded to the real numbers from a domain of [0, 1].

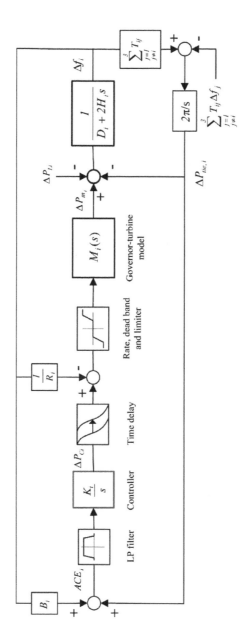

FIGURE 11.3
Frequency response model for area i.

3. Fitness values are calculated for each chromosome.

4. The fitter ones are selected as parents.

5. Some pairs of parents are selected based on the selection method and recombined to generate children.

6. Rarely, mutation is applied to the children. In other words, a few 0 bits flipped to 1, and vice versa.

7. A new population is regenerated by allowing parents and children to be together.

8. Return to step 2 and repeat the above steps until terminated conditions are satisfied.

The above procedure is schematically depicted in Figure 11.4. Following the GA application, the optimal gains for integral controllers in areas 1, 2, and 3 are obtained as 0.259, 0.278, and 0.001, respectively. The system response is examined in the presence of simultaneous 0.02 pu step load disturbances in three areas, at 2 s. The frequency deviation and the net tie-line power change in each area are shown in Figure 11.5.

It is shown that neglecting GRC, speed governor dead-band, and time delay decreases the efficiency of the designed controller in response to load disturbances in an acceptable time period.[14] The mentioned dynamics must be considered in the design of supplementary frequency control loops to eliminate their detrimental effects.

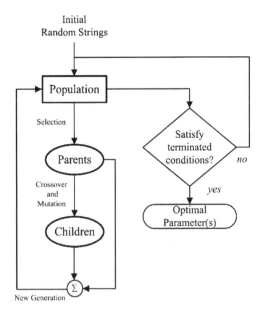

FIGURE 11.4
GA structure.

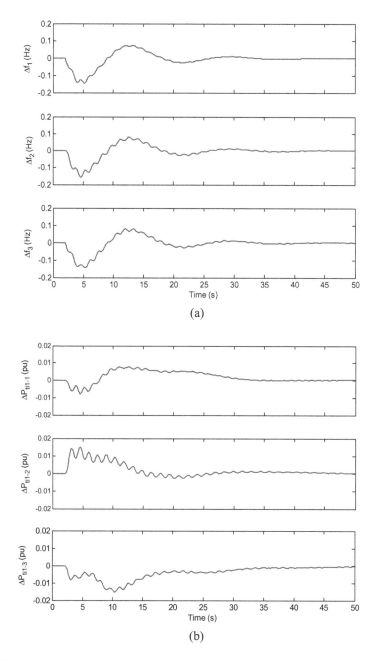

FIGURE 11.5
System response following simultaneous 0.02 pu step load increases in three areas: (a) frequency deviation and (b) tie-line power change.

11.3 Multiobjective GA

11.3.1 Multiobjective Optimization

Initially, the majority of control design problems are inherently multiobjective problems, in that there are several conflicting design objectives that need to be simultaneously achieved in the presence of determined constraints. If these synthesis objectives are analytically represented as a set of design objective functions subject to the existing constraints, the synthesis problem could be formulated as a multiobjective optimization problem.

As mentioned, in a multiobjective problem, unlike a single optimization problem, the notation of optimality is not so straightforward and obvious. Practically in most cases, the objective functions are in conflict and show different behavior, so the reduction of one objective function leads to an increase in another. Therefore, in a multiobjective optimization problem, there may not exist one solution that is best with respect to all objectives. Usually, the goal is reduced to compromise all objectives and determine a trade-off surface representing a set of nondominated solution points, known as *Pareto-optimal* solutions. A Pareto-optimal solution has the property that it is not possible to reduce any of the objective functions without increasing at least one of the other objective functions.

Unlike single-objective optimization, the solution to the multiobjective problem is not a single point, but a family of points as the (Pareto-optimal) solutions set, where any member of the set can be considered an acceptable solution. However, the choice of one solution over the other requires problem knowledge and a number of problem-related factors.[23,24]

Mathematically, a multiobjective optimization (in the form of minimization) problem can be expressed as

Minimize

$$y = f(x) = \{f_1(x), f_2(x), \dots, f_M(x)\}$$

Subject to:

$$g(x) = \{g_1(x), g_2(x), \dots, g_J(x)\} \le 0 \tag{11.2}$$

where $x = \{x_1, x_2, \dots, x_N\} \in X$ is a vector of decision variables in the decision space X, and $y = \{y_1, y_2, \dots, y_N\} \in Y$ is the objective vector in the objective space. The solution is not unique; however, one can choose a solution over the others. In the minimization case, the solution x^1 dominates x^2, or x^1 is superior to x^2, if

$$\forall i \in \{1, \dots, M\}, y(x^1) \le y(x^2) \wedge \exists i \in \{1, \dots, M\} \mid y_i(x^1) < y_i(x^2) \tag{11.3}$$

The x^1 is called a noninferior or Pareto-optimal point if any other in the feasible space of design variables does not dominate x^1. Practically, since there could be a number of Pareto-optimal solutions, and the suitability of one solution may depend on system dynamics, the environment, the designer's choice, etc., finding the center point of a Pareto-optimal solutions set may be desired.

GA is well suited for solving multioptimization problems. Several approaches have been proposed to solve multiobjective optimization problems using GAs.[15,16,25–29] The keys for finding the Pareto front among these various procedures are the Pareto-based ranking[29] and fitness-sharing[16] techniques. In the most common method, the solution is simply achieved by developing a population of Pareto-optimal or near-Pareto-optimal solutions that are nondominated. The x^i is said to be nondominated if there does not exist any x^j in the population that dominates x^i. Nondominated individuals are given the greatest fitness, and individuals that are dominated by many other individuals are given a small fitness. Using this mechanism, the population evolves toward a set of nondominated, near-Pareto-optimal individuals.[29]

In addition to finding a set of near-Pareto-optimal individuals, it is desirable that the sample of the whole Pareto-optimal set given by the set of nondominated individuals be fairly uniform. A common mechanism to ensure this is fitness sharing,[29] which works by reducing the fitness of individuals that are genetically close to each other. However, all the bits of a candidate solution bit string are not necessarily active. Thus, two individuals may have the same genotype, but different gene strings, so that it is difficult to measure the difference between two genotypes in order to implement fitness sharing. One may simply remove the multiple copies of genotypes from the population.[30]

11.3.2 Application to AGC Design

The multiobjective GA methodology is conducted to optimize the proportional-integral (PI)-based supplementary frequency control parameters in a multiarea power system. The control objectives are summarized to minimize the ACE signals in the interconnected control areas. To achieve this goal and satisfy an optimal performance, the parameters of the PI controller in each control area can be selected through minimization of the following objective function:

$$ObjFnc_i = \sum_{t=0}^{K} |ACE_{i,t}| \qquad (11.4)$$

where $ObjFnc_i$ is the objective function of control area i, K is equal to the simulation sampling time (s), and $|ACE_{i,t}|$ is the absolute value of the ACE signal for area i at time t.

Following use of the multiobjective GA optimization technique to tune the PI controllers and find the optimum values of objective functions (Equation 11.4), the fitness function (*FitFunc*) can be defined as follows:

$$FitFunc\ (\cdot) = [ObjFnc_1,\ ObjFnc_2,\ \dots,\ ObjFnc_n] \tag{11.5}$$

Each GA individual is a double vector presenting PI parameters. Since a PI controller has two gain parameters, the number of GA variables could be $N_{var\,=\,2n}$, where n is the number of control areas. The population should be considered in a matrix form with size $m \times N_{var}$; where m represents individuals.

As mentioned earlier, the population of a multiobjective GA is composed of dominated and nondominated individuals. The basic line of the algorithm is derived from a GA, where only one replacement occurs per generation. The selection phase should be done first. Initial solutions are randomly generated using a uniform random number of PI controller parameters. The crossover and mutation operators are then applied. The crossover is applied on both selected individuals, generating two children. The mutation is applied uniformly on the best individual. The best resulting individual is integrated into the population, replacing the worst-ranked individual in the population. This process is conceptually shown in Figure 11.6.

The above-described multiobjective GA is applied to the three-control area power system example used in Section 2.4. The closed-loop system response for the following simultaneous load step increase (Equation 11.6) in three areas is examined, and some results for areas 2 and 3 are shown in Figure 11.7.

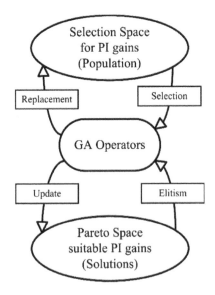

FIGURE 11.6
Multiobjective GA for tuning of PI-based supplementary frequency control parameters.

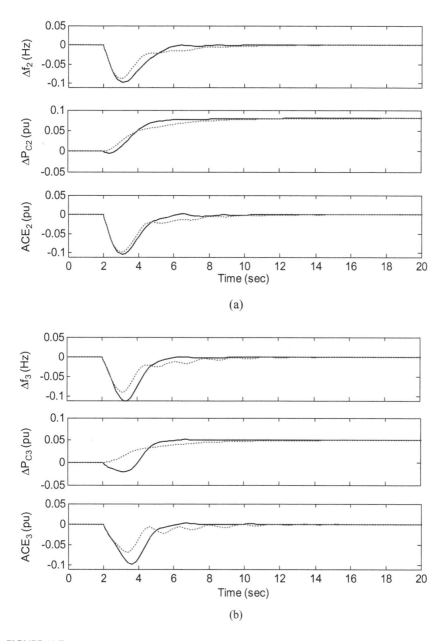

FIGURE 11.7
System responses: (a) area 2 and (b) area 3 (solid line, proposed methodology; dotted line, robust PI control).

$$\Delta P_{L1} = 100MW; \Delta P_{L2} = 80MW; \Delta P_{L3} = 50MW \qquad (11.6)$$

It has been shown from simulation results that the proposed technique is working properly, as well as the robust H$_\infty$-PI control methodology addressed in Bevrani et al.[31] Interested readers can find more time-domain simulations for various load disturbance scenarios in Daneshfar.[13]

11.4 GA for Tracking Robust Performance Index

A robust multiobjective control methodology for AGC design in a multiarea power system using the mixed H$_2$/H$_\infty$ control technique is introduced in Bevrani.[32] The AGC problem is transferred to a static output feedback (SOF) control design, and the mixed H$_2$/H$_\infty$ control is used via an iterative linear matrix inequality (ILMI) algorithm to approach a suboptimal solution for the specified design objectives.

Here, the multiobjective GA is used as a PI tuning algorithm to achieve the same robust performance as provided by ILMI-based H$_2$/H$_\infty$. In both control designs, the same controlled variables and design objectives (reducing unit wear and tear caused by equipment excursions, and addressing overshoot and number of reversals of the governor load set point signal, while area frequency and tie-line power are maintained close to specified values) are considered.

11.4.1 Mixed H$_2$/H$_\infty$

In many real-world control problems, it is desirable to follow several objectives simultaneously, such as stability, disturbance attenuation, reference tracking, and considering the practical constraints. Pure H$_\infty$ synthesis cannot adequately capture all design specifications. For instance, H$_\infty$ synthesis mainly enforces closed-loop stability and meets some constraints and limitations, while noise attenuation or regulation against random disturbances is more naturally expressed in H$_2$ synthesis terms. The mixed H$_2$/H$_\infty$ control synthesis gives a powerful multiobjective control design addressed by the LMI techniques.

A general synthesis control scheme using the mixed H$_2$/H$_\infty$ control technique is shown in Figure 11.8. $G(s)$ is a linear time-invariant system with the following state-space realization:

$$\dot{x} = Ax + B_1 w + B_2 u$$

$$z_\infty = C_\infty x + D_{\infty 1} w + D_{\infty 2} u$$

$$z_2 = C_2 x + D_{21} w + D_{22} u \qquad (11.7)$$

$$y = C_y x + D_{y1} w$$

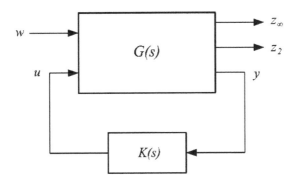

FIGURE 11.8
Mixed H_2/H_∞ control configuration.

where x is the state variable vector, w is the disturbance and other external input vector, and y is the measured output vector. The output channel z_2 is associated with the H_2 performance aspects, while the output channel z_∞ is associated with the H_∞ performance. Let $T_\infty(s)$ and $T_2(s)$ be the transfer functions from w to z_∞ and z_2, respectively.

In general, the mixed H_2/H_∞ control design method provides a dynamic output feedback (DOF) controller, $K(s)$, that minimizes the following trade-off criterion:

$$k_1 \|T_\infty(s)\|_\infty^2 + k_2 \|T_2(s)\|_2^2, \, (k_1 \geq 0, k_2 \geq 0) \tag{11.8}$$

Unfortunately, most robust control methods, such as H_2/H_∞ control design, suggest complex and high-order dynamic controllers, which are impractical for industry practices. For example, real-world AGC systems use simple PI controllers. Since a PI or proportional-integral-derivative (PID) control problem can be easily transferred to an SOF control problem,[15] one way to solve the above challenge is to use a mixed H_2/H_∞ SOF control instead of a H_2/H_∞ DOF control method. The main merit of this transformation is to use the powerful robust SOF control techniques, such as the robust mixed H_2/H_∞ SOF control, to calculate the fixed gains (PI/PID parameters), and once the SOF gain vector is obtained, the PI/PID gains are ready in hand and no additional computation is needed. In continuation, the mixed H_2/H_∞ SOF control design is briefly explained.

11.4.2 Mixed H_2/H_∞ SOF Design

The mixed H_2/H_∞ SOF control design problem can be expressed to determine an admissible SOF (pure gain vector) law K_i, belonging to a family of internally stabilizing SOF gains K_{sof},

$$u_i = K_i y_i, \, K_i \in K_{sof} \tag{11.9}$$

such that

$$\inf_{K_i \in K_{sof}} \left\| T_2(s) \right\|_2 \text{ subject to } \left\| T_\infty(s) \right\|_\infty < 1 \qquad (11.10)$$

The optimization problem given in Equation 11.10 defines a robust performance synthesis problem, where the H_2 norm is chosen as the performance measure. There are some proper lemmas giving the necessary and sufficient condition for the existence of a solution for the above optimization problem to meet the following performance criteria:

$$\left\| T_2(s) \right\|_2 < \gamma_2, \left\| T_\infty(s) \right\|_\infty < \gamma_\infty \qquad (11.11)$$

where γ_2 and γ_∞ are H_2 and H_∞ robust performance indices, respectively.

It is notable that the H_∞ and H_2/H_∞ SOF reformulation generally leads to bilinear matrix inequalities that are nonconvex. This kind of problem is usually solved by an iterative algorithm that may not converge to an optimal solution. An ILMI algorithm is introduced in Bevrani[32] to get a suboptimal solution for the above optimization problem. The proposed algorithm searches the desired suboptimal H_2/H_∞ SOF controller K_i within a family of H_2 stabilizing controllers K_{sof} such that

$$\left| \gamma_2^* - \gamma_2 \right| < \varepsilon, \gamma_\infty = \left\| T_{z_{\infty i} \, v_{1i}} \right\|_\infty < 1 \qquad (11.12)$$

where ε is a small real positive number, γ_2^* is H_2 performance corresponding to H_2/H_∞ SOF controller K_i, and γ_2 is the optimal H_2 performance index that can be obtained from the application of the standard H_2/H_∞ DOF control.

11.4.3 AGC Synthesis Using GA-Based Robust Performance Tracking

The design of a robust SOF controller based on a H_2/H_∞ control was discussed in the previous section. Now, the application of GA for getting pure gains (SOF) is presented to achieve the same robust performances (Equation 11.11). Here, like in the H_2/H_∞ control scheme shown in Figure 11.8, the optimization objective is to minimize the effects of disturbances (w) on the controlled variables (z_∞ and z_2). This objective can be summarized as

$$Min \; \gamma_2 = \left\| T_2(s) \right\|_2 \; Subject \; to \; \gamma_\infty = \left\| T_\infty(s) \right\|_\infty \qquad (11.13)$$

such that the resulting performance indices (γ_2^*, γ_∞^*) satisfy $\left| \gamma_2^* - \gamma_2^{opt} \right| < \varepsilon$ and $\gamma_\infty^* < 1$. Here, ε is a small real positive number, γ_2^* and γ_∞^* are H_2 performance and H_∞ performance corresponding to the obtained controller K_i from the GA optimization algorithm, and γ_2^{opt} is the optimal H_2 performance index

TABLE 11.1

PI Parameters and Optimal Performance Index

Design Technique	ILMI-Based H_2/H_∞			Multiobjective GA		
Areas	Area 1	Area 2	Area 3	Area 1	Area 2	Area 3
K_{Pi}	−2.00E-04	−4.80E-03	−2.50E-03	−1.00E-04	−0.0235	−1.00E-04
K_{Ii}	−0.3908	−0.4406	−0.4207	−0.2309	−0.2541	−0.2544
γ^*_{2i}	1.0976	1.0345	1.0336	1.0371	0.9694	0.9807
$\gamma^*_{\infty i}$	0.3920	0.2950	0.3498	0.3619	0.2950	0.3497

that can be achieved from the application of standard H_2/H_∞ dynamic output feedback control. In order to calculate γ_2^{opt}, one may simply use the *hinfmix* function in MATLAB based on the LMI control toolbox.[33]

The proposed control technique is applied to the three-control area power system given in Section 2.4. In the proposed approach, the GA is employed as an optimization engine to produce the PI controllers in the supplementary frequency control loops with performance indices near the optimal ones. The obtained control parameters and performance indices are shown in Table 11.1. The indices are comparable to the results given by the proposed ILMI algorithm. For the problem at hand, the guaranteed optimal H_2 performance indices (γ_2^{opt}) for areas 1, 2, and 3 are calculated as 1.070, 1.03, and 1.031, respectively.

Figure 11.9 shows the closed-loop response (frequency deviation, area control error, and control action signals) for areas 1 and 3, in the presence of simultaneous 0.1 pu step load disturbances, and a 20% decrease in the inertia constant and damping coefficient as uncertainties in all areas. The performance of the closed-loop system using GA-based H_2/H_∞ PI controllers is also compared with that of the ILMI-based H_2/H_∞ PI control design. Simulation results demonstrate that the proposed GA-based PI controllers track the load fluctuations and meet robustness for a wide range of load disturbances, as well as ILMI-based PI controllers.

11.5 GA in Learning Process

GAs belong to a class of adaptive general purpose methods, for machine learning, as well as optimization, based on the principles of population genetics and natural evolution. A GA learns by evaluating its knowledge structures using the fitness function, and forming new ones to replace the previous generation by breeding from more successful individuals in the population using the crossover and mutation operators.

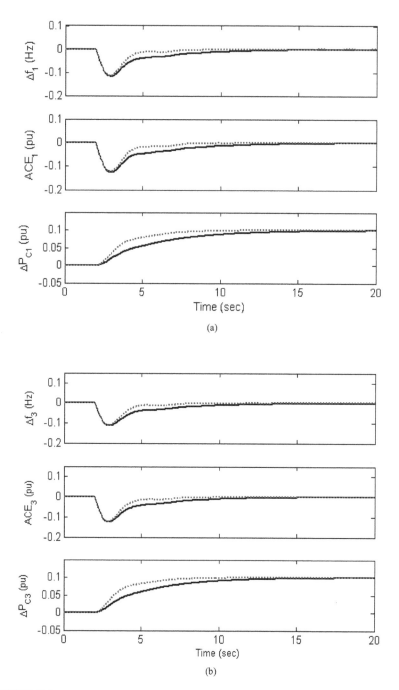

FIGURE 11.9
Closed-loop system response: (a) area 1 and (b) area 3 (solid, GA; dotted, ILMI).

In Section 7.3, the application of GA to find suitable state and action variables in the reinforcement learning (RL) process for a multiagent RL-based AGC design was described. In the present section, as another example, it is shown that the GA can also be effectively used to provide suitable training data in a multiagent Bayesian-network (BN)-based AGC (see Chapter 8).

11.5.1 GA for Finding Training Data in a BN-Based AGC Design

As mentioned in Chapter 8, the basic structure of a graphical model is built based on prior knowledge of the system. Here, for simplicity, assume that frequency deviation and tie-line power change are the most important AGC variables, regarding the least dependency to the model parameters and the maximum effectiveness on the system frequency. In this case, the graphical model for the BN-based AGC scheme can be considered as shown in Figure 11.10, and the posterior probability that can be found is $p(\Delta P_c | \Delta P_{tie}, \Delta f)$.

After determining the most worthwhile subset of observations ($\Delta P_{tie}, \Delta f$), in the next phase of BN construction, a directed acyclic graph that encodes assertion of conditional independence is built. It includes the problem random variables, conditional probability distribution nodes, and dependency nodes.

As mentioned in Chapter 8, in the next step of BN construction, i.e., parameter learning, the local conditional probability distributions $p(x_i | pa_i)$ are computed from the training data. According to the graphical model (Figure 11.10), probability and conditional probability distributions for this problem are $p(\Delta f)$, $p(\Delta P_{tie})$, and $p(\Delta P_C | \Delta f, \Delta P_{tie})$. To calculate the above probabilities, suitable training data are needed. In the learning phase, to find the conditional probabilities related to the graphical model variables, the training data can be used in a proper software environment.[34]

Here, GA is applied to obtain a set of training data (ΔP_{tie}, Δf, ΔP_C) as follows: In an offline procedure, a simulation is run with an initial ΔP_C vector provided by GA for a specific operating condition. Then, the appropriate ΔP_C is evaluated based on the calculated ACE signal. Each GA's individual ΔP_C

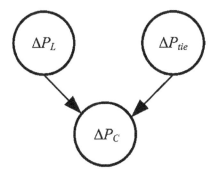

FIGURE 11.10
The graphical model of area *i*.

is a double vector (population type) with n_v variables in the range [0 1] (the number of variables can be considered the same as the number of simulation seconds). In the simulation stage, the vector's elements should be scaled to a valid ΔP_C change for that area: $[\Delta P_{Cmin}\ \Delta P_{Cmax}]$. The ΔP_{Cmax} is the possible maximum control action change to use for an AGC cycle, and similarly, ΔP_{Cmin} is possible minimum change that can be applied to the governor set point.

11.5.2 Application Example

Consider the three-control area power system (used in the previous section) again. Here, the start population size is fixed at thirty individuals and run for one hundred generations. Figure 11.11 shows the results of running the proposed GA for area 1. To examine the individual's eligibility (fitness), ΔP_C values should be scaled according to the specified range for the control area. After scaling and finding the corresponding ΔP_C, the simulation is run for a given load disturbance (a signal with one hundred instances) or the scaled ΔP_C (with 100 s). The individual fitness is proportional to the average distances of the resulting ACE signal instances from zero, after 100 s of simulation. Finally, the individuals with higher fitness are the best ones, and the resulting set of $(\Delta P_{tie}, \Delta f, \Delta P_C)$ provides a row of the training data matrix.

As explained in Chapter 8, this large training data matrix is partly complete, and it can be used for parameter learning issues in the power system

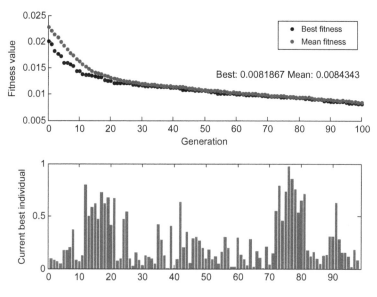

FIGURE 11.11
Result of running GA for area 1.

for other disturbance scenarios and operating conditions. Here, the Bayesian networks toolbox (BNT)[34] is used as suitable software for simulation purposes. After providing the training set, the training data for each area are separately supplied to the BNT. The BNT uses the data and the parameter learning phase for each control area parameter, and determines the associated prior and conditional probability distributions $p(\Delta f)$, $p(\Delta P_{tie})$.

After completing the learning phase, we are ready to run the AGC system online, and the proposed model uses the inference phase to find an appropriate control action signal (ΔP_C) for each control area as follows: at each simulation time step, corresponding control agents get the input parameters (ΔP_{tie}, Δf) of the model and digitize them for the BNT (the BNT does not work with continuous values). The BNT finds the posterior probability distribution $p(\Delta P_C | \Delta P_{tie}, \Delta f)$ for each area, and then the control agent finds the maximum posterior probability distribution from the return set and provides the most probable evidence ΔP_C.

The response of the closed-loop system (for areas 2 and 3) in the presence of the disturbance scenario given in Equation 11.6 is shown in Figure 11.12. The performance of the proposed GA-based multiagent BN is compared with that of the robust PI control design presented in Bevrani and coworker.[31]

11.6 Summary

The GAs are emerging as powerful alternatives to the traditional optimization methods. As GAs are inherently adaptive, they can effectively converge to near optimal solutions in many applications, and therefore they have been used to solve a number of complex problems over the years. A GA performs the task of optimization by starting with a random population of values, and producing new generations of improved values that combine the values with best fitness from previous populations. GAs can efficiently handle highly nonlinear and noisy cost functions, and therefore they can be considered powerful optimization tools for real-world complex dynamic systems.

This chapter started with an introduction on GA algorithms and their applications in control systems. Then, several methodologies were presented for the GA-based AGC design problem. GAs are successfully used for the AGC system with different strategies, in the form of tuning of controller parameters, solving of multiobjective optimization problems, tracking of robust performance indices, and improving learning algorithms. The proposed design methodologies are illustrated by suitable examples. In most cases, the results are compared with those of recent robust control designs.

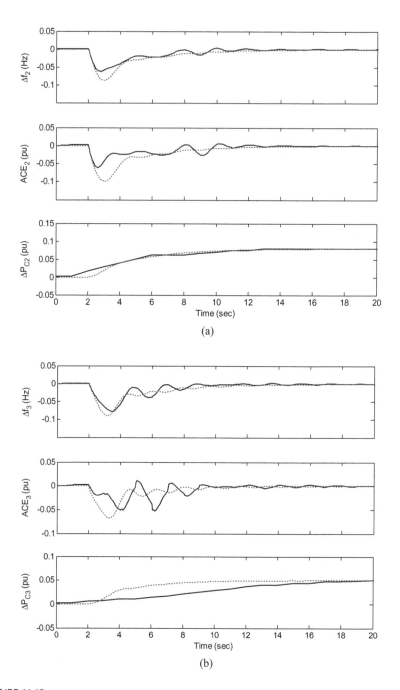

FIGURE 11.12
System response: (a) area 2 and (b) area 3 (solid line, proposed method; dotted line, robust
PI controller).

References

1. D. Rerkpreedapong, A. Hasanovic, A. Feliachi. 2003. Robust load frequency control using genetic algorithms and linear matrix inequalities. *IEEE Trans. Power Syst.* 18(2):855–61.
2. Y. L. Abdel-Magid, M. M. Dawoud. 1996. Optimal AGC tuning with genetic algorithms. *Elect. Power Syst. Res.* 38(3):231–38.
3. A. Abdennour. 2002. Adaptive optimal gain scheduling for the load frequency control problem. *Elect. Power Components Syst.* 30(1):45–56.
4. S. K. Aditya, D. Das. 2003. Design of load frequency controllers using genetic algorithm for two area interconnected hydro power system. *Elect. Power Components Syst.* 31(1):81–94.
5. C. S. Chang, W. Fu, F. Wen. 1998. Load frequency controller using genetic algorithm based fuzzy gain scheduling of PI controller. *Elect. Machines Power Syst.* 26:39–52.
6. Z. M. Al-Hamouz, H. N. Al-Duwaish. 2000. A new load frequency variable structure controller using genetic algorithm. *Elect. Power Syst. Res.* 55:1–6.
7. A. Huddar, P. S. Kulkarni. 2008. A robust method of tuning the feedback gains of a variable structure load frequency controller using genetic algorithm optimization. *Elect. Power Components Syst.* 36:1351–68.
8. P. Bhatt, R. Roy, S. P. Ghoshal. 2010. Optimized multi area AGC simulation in restructured power systems. *Elect. Power Energy Syst.* 32:311–22.
9. A. Demirorem, S. Kent, T. Gunel. 2002. A genetic approach to the optimization of automatic generation control parameters for power systems. *Eur. Trans. Elect. Power* 12(4):275–81.
10. P. Bhatt, R. Roy, S. P. Ghoshal. 2010. GA/particle swarm intelligence based optimization of two specific varieties of controller devices applied to two-area multi-units automatic generation control. *Elect. Power Energy Syst.* 32:299–310.
11. K. Vrdoljak, N. Peric, I. Petrovic. 2010. Sliding mode based load-frequency control in power systems. *Elect. Power Syst. Res.* 80:514–27.
12. F. Daneshfar, H. Bevrani. 2010. Load-frequency control: A GA-based multi-agent reinforcement learning. *IET Gener. Transm. Distrib.* 4(1):13–26.
13. F. Daneshfar. 2009. Automatic generation control using multi-agent systems. MSc dissertation, Department of Electrical and Computer Engineering, University of Kurdistan, Sanandaj, Iran.
14. H. Golpira, H. Bevrani. 2010. Application of GA optimization for automatic generation control in realistic interconnected power systems. In *Proceedings of International Conference on Modeling, Identification and Control*, Okayama, Japan, CD-ROM.
15. H. Bevrani, T. Hiyama. 2007. Multiobjective PI/PID control design using an iterative linear matrix inequalities algorithm. *Int. J. Control Automation Syst.* 5(2):117–127.
16. D. E. Goldberg. 1989. *Genetic algorithms in search, optimization and machine learning*. Reading, MA: Addison-Wesley.
17. L. Davis. 1991. *Handbook of genetic algorithms*. New York: Van Nostrand.
18. K. F. Man, K. S. Tag. 1997. Genetic algorithms for control and signal processing. In *Proceedings of IEEE International Conference on Industrial Electronics, Control and Instrumentation-IECON*, New Orleans, LA, USA, vol. 4, pp. 1541–55.

19. J. Hu, E. Goodman. 2005. Topological synthesis of robust dynamic systems by sustainable genetic programming. In *Genetic programming theory and practice II*, ed. U. M. O'Reilly, T. Yu, R. Riolo, B. Wozel, New York: Springer, pp. 143–58.
20. B. Forouraghi. 2000. A genetic algorithm for multiobjective robust design. *Appl. Intelligence* 12:151–61.
21. M. O. Odetayo, D. Dasgupta. 1995. Controlling a dynamic physical system using genetic-based learning methods. In *Practical handbook of genetic algorithms—New frontiers*, ed. L. Chambers. Vol. II. Boca Raton, FL: CRC Press.
22. H. Bevrani. 2009. Real power compensation and frequency control. In *Robust power system frequency control*, pp. 15–38. New York: Springer.
23. J. L. Cohon. 1978. *Multiobjective programming and planning*. New York: Academic Press.
24. A. Osyczka. 1985. Multicriteria optimization for engineering design. In *Design optimization*, ed. J. S. Gero, 193–227. New York: Academic Press.
25. J. D. Schaer. 1984. Some experiments in machine learning using vector evaluated genetic algorithms. Unpublished doctoral dissertation, Vanderbilt University, Nashville, TN.
26. N. Srinivas, K. Deb. 1995. Multiobjective optimization using non-dominated sorting in genetic algorithm. *Evol. Comput.* 2(3):221–48.
27. H. Tamaki, H. Kita, S. Kobayashi. Multi-objective optimization by genetic algorithms: A review. In *Proceedings of 1996 IEEE International Conference on Evolutionary Computation*, Nagoya, Japan, pp. 517–22.
28. C. Poloni, et al. 1995. *Hybrid GA for multi objective aerodynamic shape optimization, genetic algorithms in engineering and computer science*, ed. C. Winter, et al., 397–416. Chichester, UK: Wiley.
29. C. M. Fonseca, P. J. Fleming. 1995. Multiobjective optimization and multiple constraint handling with evolutionary algorithms. Part I. A unified formulation. *IEEE Trans. Syst. Man Cyber. A* 28(I):26–37.
30. J. F. Whidborne, R. S. H. Istepanian. 2001. Genetic algorithm approach to designing finite-precision controller structures. *IEE Proc. Control Theory Appl.* 148(5):377–82.
31. H. Bevrani, Y. Mitani, K. Tsuji. 2004. Robust decentralised load-frequency control using an iterative linear matrix inequalities algorithm. *IEE Proc. Gener. Transm. Distrib.* 3(151):347–54.
32. H. Bevrani. 2009. Multi-objective control-based frequency regulation. In *Robust power system frequency control*, 103–22. New York: Springer.
33. P. Gahinet, A. Nemirovski, A. J. Laub, M. Chilali. 1995. *LMI control toolbox*. Natick, MA: The MathWorks.
34. K. Murphy. 2001. The Bayes net toolbox for Matlab. *Comput. Sci. Stat.* 33: 1–20.

12

Frequency Regulation in Isolated Systems with Dispersed Power Sources

Numerous new distributed power generation technologies, such as the photovoltaic (PV) generation, the wind turbine generation, the micro gas turbine generation, and the energy storage devices, are currently available to offer integrated performance and flexibility for the power consumers.[1-4] Frequency regulation in interconnected networks is one of the main control challenges posed by emerging new uncertainties and numerous distributed generators, including renewable energy sources in a modern power system.[5] Significant interconnection frequency deviations due to distributed power fluctuations can cause under- or overfrequency relaying and disconnect some loads and generations. Under unfavorable conditions, this may result in a cascading failure and system collapse.[6]

This chapter presents a multiagent-based automatic generation control (AGC) scheme for isolated power systems[7-9] with dispersed power sources such as PV units, wind generation units, diesel generation units, and energy capacitor systems (ECS)[10-12] for the energy storage. The ECS consists of electrical double-layer capacitors. The power generation from the PV units and also from the wind generation units depends on environmental factors, such as the solar insolation and wind velocity; therefore, complete regulation of the power from these units is quite difficult.

Contribution of ECS units to the frequency regulation in coordination with conventional AGC participant generating units was addressed in Chapter 10, and the application of multiagent systems in AGC synthesis was discussed in Chapters 7 and 8. In this chapter, the ECS is coordinated with the diesel units to propose a new multiagent-based AGC scheme.

As mentioned in Chapter 10, since the ECS units are able to provide a fast charging/discharging operation, the variations of power from the wind turbine units and also from the PV units can be absorbed through the charging or discharging operation of the ECS units. In addition, the variation of power consumption at the variable load can also be absorbed through the charging/discharging operation of the ECS units. A small-sized ECS is considered in this study; therefore, the continuous charging or discharging operation is not available on the ECS because of its restricted capacity. To overcome this situation, the regulation power on the diesel units is inevitable to keep the stored energy of the ECS in a proper range for the continuous AGC operation on the ECS. In the proposed AGC scheme, the ECS provides the main function of AGC, while the diesel units provide a supplementary function of the AGC system. Namely, a coordinated AGC between the ECS and the diesel units has been proposed for balancing the total power generation and the total power demand in the isolated power systems. The proposed multiagent system consists of three types of agents: monitoring agents for the distribution of required information through the computer network, control agents for the charging/discharging operation on the ECS and also for the power regulation on the diesel units, and finally, a supervisor agent for the coordination between the ECS and diesel units. Experimental studies in a power system laboratory have been performed to demonstrate the efficiency of the proposed multiagent-based AGC scheme.

12.1 Configuration of Multiagent-Based AGC System

In the proposed AGC scheme, the system frequency monitored on a diesel unit is regulated to its nominal frequency by balancing the total generation and the total power consumption in the isolated power system. Namely, the AGC function can be achieved through the charging/discharging operation of the ECS in coordination with the produced regulation power from the available diesel units.

Figure 12.1 illustrates the configuration of the proposed multiagent-based AGC system for isolated power systems with dispersed power sources. This configuration, which will be described later, is similar to the agent-based frequency regulation scheme presented in Figure 3.8.

12.1.1 Conventional AGC on Diesel Unit

In the conventional scheme, the diesel units are utilized for the AGC system to regulate the power generation following the monitored frequency deviation on itself. Figure 12.2 illustrates the configuration of the AGC system based on the flat frequency control (FFC) with a proportional-integral (PI) control loop. Here, Δf and ΔP_C represent the measured frequency deviation on the diesel unit and the provided signal for the output setting of the diesel unit.

12.1.2 Coordinated AGC on the ECS and Diesel Unit

In this study, a new AGC scheme has been proposed considering the coordination between the ECS and the diesel units. The basic configuration of the proposed feedback control system for the ECS on the supervisor agent is shown in Figure 12.3. Here, T_{delay} and PS_{ECS} represent the communication time delay[13] and the control signal for the output setting of the ECS unit. The configuration, except the time delay block, is the same as that in Figure 12.2,

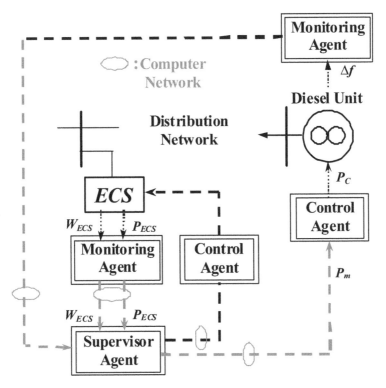

FIGURE 12.1
Configuration of multiagent-based AGC system.

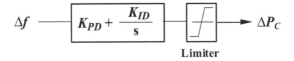

FIGURE 12.2
Configuration of AGC on diesel unit.

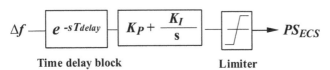

FIGURE 12.3
Feedback control system for ECS on the supervisor agent.

where the diesel units are utilized for the AGC. Applying the control signal PS_{ECS} from the mentioned loop provides an appropriate charging/discharging operation on the ECS for the frequency regulation purpose.

Because of the specific feature of the ECS dynamics, it is possible to achieve the fast charging/discharging operation in an ECS unit. Therefore, the variations

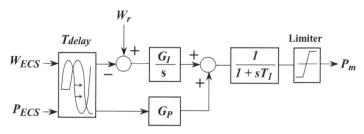

FIGURE 12.4
Coordinated AGC system for diesel unit on the supervisor agent.

of power generation from the wind turbine and PV units, as well as the variation of demand power on the variable loads, can be efficiently absorbed through the charging/discharging operation of the ECS unit. A small-sized ECS is considered in this study. Therefore, similar to the proposed frequency regulation in Chapter 10, an additional regulation power (from the diesel units) is required to keep the stored energy level of the ECS in a proper range.

Figure 12.4 illustrates the configuration of the coordinated controller for the diesel unit on the supervisor agent. In this figure, W_r and W_{ECS} are the target and measured (current) stored energies. P_{ECS} and P_m represent the regulation powers provided by the ECS and diesel units, respectively. In the proposed AGC scheme, the ECS provides the main function of the AGC, and the diesel units provide a supplementary function to support the charging or discharging operation on the ECS unit. Namely, a coordinated AGC between the ECS and the diesel units has been performed to balance the power demand and the total power generation.

In order to realize the proposed coordinated AGC scheme, a multiagent system has been utilized, as shown in Figure 12.1. The required AGC performance is achieved through the charging/discharging operation on the ECS following the monitored frequency deviation Δf on the diesel unit. As shown in Figure 12.1, three different types of agents are specified in the proposed multiagent-based AGC system: *monitoring agents* for the distribution of required information through the computer network, *control agents* for the charging/discharging operation on the ECS and also for the power regulation on the diesel units, and finally, a *supervisor agent* for the coordination between the ECS and the diesel units. These three agent types can communicate with each other through a secure computer network to achieve a desirable AGC performance.

12.2 Configuration of Laboratory System

The simplified single line diagram of the studied laboratory system is shown in Figure 12.5. Figure 12.6 shows an overview of the experimental laboratory

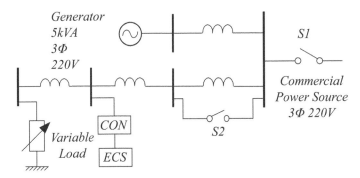

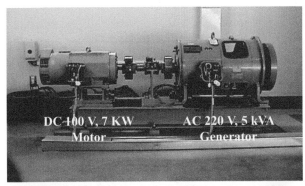

FIGURE 12.5
Single line diagram of the laboratory system.

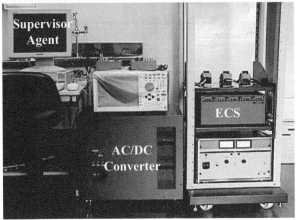

FIGURE 12.6
Overview of the experimental laboratory system.

system. The laboratory system consists of a 5 kVA generator driven by a DC motor representing the diesel unit, a 70 Wh ECS with the maximum charging/discharging power of 3 kW, a variable load, and the transmission line modules. The variations of power generation from the PV and the wind turbine units are represented by the variations of power consumption on the variable load. During the start-up process, the laboratory system is connected to the external commercial power source, i.e., the switch S_1 is closed. Following the start-up process, the switch S_1 is opened to change the study system to an isolated power system.

12.3 Experimental Results

The performance of the proposed AGC scheme is examined in the presence of various load disturbance scenarios. The tuning of control parameters was performed for the step load change scenario 1 shown in Figures 12.7 and 12.8.

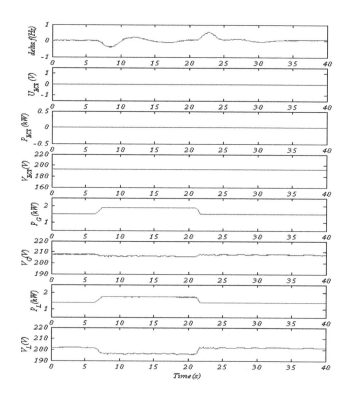

FIGURE 12.7
Conventional AGC for step load change scenario 1.

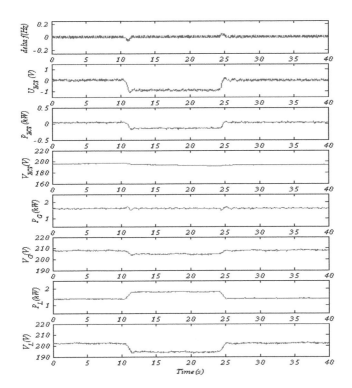

FIGURE 12.8
Proposed AGC for step load change scenario 1 (ECS controller: $K_P = 4$, $K_I = 10$).

The tuned parameters for the ECS controller are as follows: $K_P = 4$ and $K_I = 10$. In this case, the step load change is applied just for a specific range of time. Therefore, the integration of the discharging power did not reach its critical level. Namely, the coordination from the diesel unit is not required in this case. However, in the cases of the step load change scenarios 2 and 3, the step load changes are not cleared; therefore, the coordination from the diesel unit was inevitable to keep the stored energy of the ECS within a prespecified range. The diesel unit cannot follow the fast random load change because of its response speed; therefore, the coordination is not considered for the random load change. For the large disturbance caused by the line switching S_2, coordination is not necessary because any load change is not applied to the laboratory system.

Typical experimental results are illustrated in Figures 12.7 to 12.14. In these figures, the frequency deviation Δf (Hz) monitored on the generator (representing the diesel unit), the control signal U_{ECS} (V) for the charging or discharging operation on the ECS, the charging/discharging power P_{ECS} (kW) on the ECS, the ECS terminal voltage V_{ECS} (V), the generator power P_G (kW), the generator terminal voltage V_G (V), the power consumption P_L (kW) on the

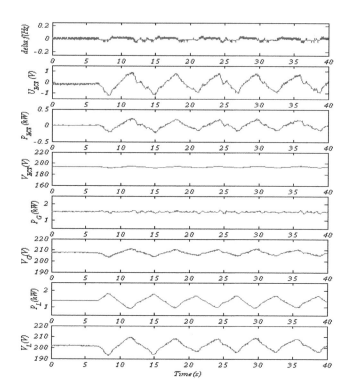

FIGURE 12.9
Proposed AGC for periodical load change (scenario 4).

variable load, and the terminal voltage V_L (V) of the variable load are illustrated from the top to the bottom.

The averaged frequency deviation and the maximum frequency deviation are shown in Table 12.1 for both the conventional AGC and the proposed multiagent-based AGC under different types of load changes, and the large disturbance given by the line switching. As clearly indicated in Table 12.1, the AGC performance is highly improved by applying the proposed multiagent-based AGC scheme. The estimated time delay is around 70 ms during the experiments. For the larger time delay, the compensation is inevitable to maintain better control performance.

The stored energy on the ECS is easily monitored by the voltage V_{ECS} (V) measured at the ECS terminal. During the experiments, the operation range of the ECS terminal voltage V_{ECS} (V) is specified from 140 to 240 V. As shown in Figures 12.10 and 12.11, the ECS terminal voltage was kept almost constant by the coordination from the generator representing the diesel unit.

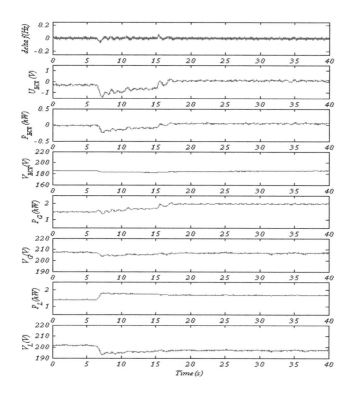

FIGURE 12.10
Proposed AGC for step load change scenario 2.

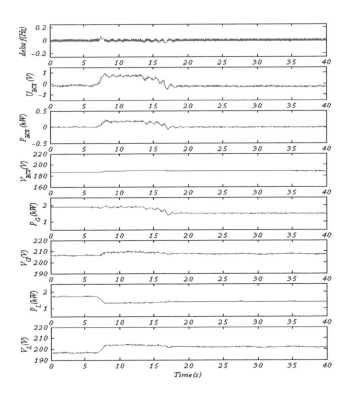

FIGURE 12.11
Proposed AGC for step load change scenario 3.

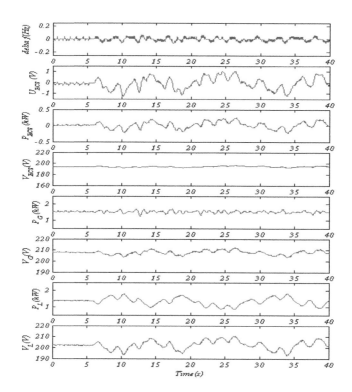

FIGURE 12.12
Proposed AGC for random load change (scenario 5).

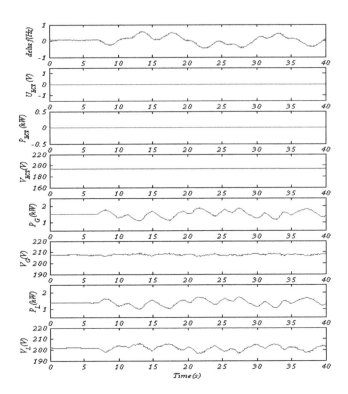

FIGURE 12.13
Random load change without control (scenario 5).

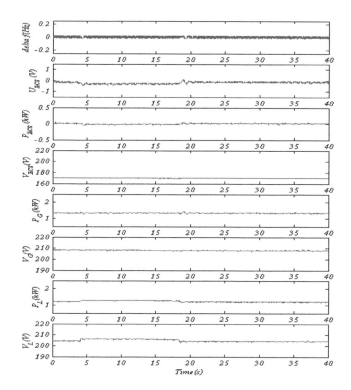

FIGURE 12.14
Proposed AGC under large disturbance without load change (scenario 6).

TABLE 12.1

Averaged and Maximum Frequency Deviation under
Different Load Change Scenarios

(a) Step Load Change Scenario 1

AGC Scheme	Δf_{ave} (Hz)	Δf_{max} (Hz)
Conventional	0.2632	0.8019
Multiagent	0.0105	0.0676

(b) Step Load Change Scenario 2

AGC Scheme	Δf_{ave} (Hz)	Δf_{max} (Hz)
Conventional	0.0518	0.3957
Multiagent	0.0087	0.0695

(c) Step Load Change Scenario 3

AGC Scheme	Δf_{ave} (Hz)	Δf_{max} (Hz)
Conventional	0.0623	0.4771
Multiagent	0.0085	0.0609

(d) Periodical Load Change (Scenario 3)

AGC Scheme	Δf_{ave} (Hz)	Δf_{max} (Hz)
Conventional	0.2571	0.7018
Multiagent	0.0161	0.0560

(e) Random Load Change (Scenario 5)

AGC Scheme	Δf_{ave} (Hz)	Δf_{max} (Hz)
Conventional	0.1986	0.5684
Multiagent	0.0187	0.0667

*(f) Line Switching Disturbance without Load Change
   (Scenario 6)*

AGC Scheme	Δf_{ave} (Hz)	Δf_{max} (Hz)
Conventional	0.0499	0.1013
Multiagent	0.0094	0.0276

12.4 Summary

An intelligent multiagent-based automatic generation control scheme for a
power system case study with dispersed power sources such as photovoltaic, wind generation, diesel generation, and an energy capacitor system is
proposed. An experimental study is used to demonstrate the capability of
the proposed control structure. The experimental results clearly indicate the
advantages of the proposed control scheme in comparison to the conventional AGC framework.

References

1. F. Bonnano, A. Consoli, A. Raciti, B. Morgana, U. Nocera. 1999. Transient analysis of integrated diesel-wind-photovoltaic generation systems. *IEEE Trans. Energy Conversion* 14(2):232–38.
2. R. Ramakumar, L. Abouzahr, K. Krishnan, K. Ashenayi. 1995. Design scenarios for integrated renewable energy systems. *IEEE Trans. Energy Conversion* 10(4):736–46.
3. G. S. Stavrakakis, G. N. Kariniotakis. 1995. A general simulation algorithm for the accurate assessment of isolated diesel-wind turbines systems interaction. Part I. A general multimachine power system model. *IEEE Trans. Energy Conversion* 10(3):577–81.
4. G. S. Stavrakakis, G. N. Kariniotakis. 1995. A general simulation algorithm for the accurate assessment of isolated diesel-wind turbines systems interaction. Part II. Implementation of the algorithm and case studies with induction generators. *IEEE Trans. Energy Conversion* 10(3):584–90.
5. H. Bevrani, F. Daneshfar, P. R. Daneshmand. 2010. Intelligent power system frequency regulation concerning the integration of wind power units. In *Wind power systems: Applications of computational intelligence,* ed. L. F. Wang, C. Singh, A. Kusiak, 407–37. Springer Book Series on Green Energy and Technology. Heidelberg: Springer-Verlag.
6. H. Bevrani. 2009. *Robust power system frequency control.* New York: Springer.
7. T. Hiyama, K. Tomsovic, M. Yoshimoto, Y. Hori. 2001. Modeling and simulation of distributed power sources. In *Proceedings of the IPEC 2001*, Singapore, vol. 2, pp. 634–38.
8. T. Hiyama, D. Zuo, T. Funabashi. 2002. Automatic generation control of stand alone power system with energy capacitor system. In *Proceedings of the Fifth International Conference on Power System Management and Control (PSMC 2002)*, London, pp. 59–64.
9. T. Hiyama, D. Zuo, T. Funabashi. 2002. Multi-agent based automatic generation control of isolated stand alone power system. In *Proceedings of 2002 International Conference on Power System Technology (PowerCon 2002)*, Kunming, China, vol. 1, pp. 139–43.
10. M. Okamoto. 1995. A basic study on power storage capacitor systems. *Trans. IEE Jpn.* 115–B(5).
11. M. Ohshima, M. Shimizu, M. Shimizu, M. Yamagishi, M. Okamura. 1998. Novel utility-interactive electrical energy storage system by electrical double layer capacitors and an error tracking mode PWM converter. *Trans. IEE Jpn.* 118-D(12).
12. T. Hiyama, D. Ueno, S. Yamashiro, M. Yamagishi, M. Shimizu. 2000. Fuzzy logic switching control for electrical double-layer energy capacitor system for stability enhancement. In *Proceedings of the IEEE PES 2000 Summer Meeting*, Seattle, WA, USA, vol. 4, pp. 2002–7.
13. H. Bevrani, T. Hiyama. 2009. On robust load-frequency regulation with time delays: Design and real-time implementation. *IEEE Trans. Energy Conversion* 24(1):292–300.

Index